Modern Logic and Quantum Mechanics

Modern Logic and Quantum Mechanics

Rachel Wallace Garden

Adam Hilger Ltd, Bristol

British Library Cataloguing in Publication Data

Garden, Rachel Wallace
    Modern Logic and Quantum Mechanics
    1. Quantum theory   2. Physics—Philosophy
    I. Title
    530.1′2   QC174.13

    ISBN 0-85274-457-9

Published by Adam Hilger Ltd, Techno House, Redcliffe Way, Bristol, BS1 6NX.
The Adam Hilger book-publishing imprint is owned by The Institute of Physics.

Filmset in Baskerville by Mid-County Press, London, SW15.
Printed in Great Britain by J W Arrowsmith Ltd, Bristol.

To Peter, Emily, Daniel and Sally

Contents

Acknowledgments

I should like to thank Professor Jeff Bub from the University of Western Ontario, and Professor Cliff Hooker, formerly of UWO, now at the University of Newcastle, Australia, for reading some parts of this book in draft and making useful comments. Professor William Harper of UWO also banished some barbarisms at the last moment, and was encouraging throughout. I want also to thank the Logic Group at Victoria University of Wellington, New Zealand, for listening to parts of chapter 3 in rudimentary form. I am grateful to the International Federation of University Women, the provincial government of Ontario, and the University of Western Ontario for support in the early days of this research.

Perhaps most of all I am grateful to my friends in North Wales who helped me find time to write this book. I want especially to thank my husband Peter Garden, and also my sister Cath Wallace, who helped me in spite of important commitments of her own. I want also to thank my parents, Charlotte and Lindsay Wallace, for their support and encouragement.

Introduction

In this book we shall apply the methods of modern logic to one of the most interesting problems in the philosophy of physics. This is the problem of 'interpreting' quantum mechanics. Quantum theory has been the subject of argument ever since it was developed early this century. For example, Einstein and Bohr had celebrated disagreements about how it should properly be understood, and they always remained utterly opposed to each other's views. The important questions remain open. Much has been written in the last fifty years about the conceptual problems of quantum mechanics, but very little has been resolved.

This book is not in the tradition of the 'quantum logic' school, a view which is discussed in detail and rejected here. According to the 'quantum logicians' we use a system of reasoning in quantum mechanics which is quite unlike the ordinary logic we use for other things. The arguments for this interpretation of quantum mechanics will be examined in detail here (in §7.3), but it will not be accepted.

The first chapter gives a brief introduction to quantum mechanics. The mathematical structure of quantum theories is first examined, since this structure lies at the heart of the conceptual problems. This is a self-contained discussion, but is bound to be a little difficult for anyone unfamiliar with the mathematical terms. In the second section of chapter 1 a less technical discussion is given of some well-known quantum 'paradoxes', and some of the solutions which have been proposed. The debate between Einstein and Bohr is also discussed.

Chapters 2 and 3 analyse the propositional logic of mechanical theories. Chapter 2 introduces modern logic to those new to the subject. In the first section we show how a propositional logic is defined, using classical bivalent logic as an illustration. In the second section we introduce the modern algebraic methods, including all the fundamental definitions. Chapter 3 applies the methods of chapter 2 to mechanical theories, generating the propositional logic of mechanics from first principles.

Chapters 4, 5 and 6 develop a logical foundation for probability assignments. In chapter 4, we discuss the states of a theory and the logical role of measurement, using differences in the structure of states to distinguish classical theories from non-classical ones, such as quantum mechanics. The first section of chapter 5 gives an introduction to standard probability theory. In the following sections, we see the traditional way this is used to analyse classical mechanics, and the problems encountered when we try to apply it in the same way to quantum theories. In chapter 6, a new analysis is given of logic and probabilities. We show how a state of mechanics generates conditional probabilities. In classical theories these have all the familiar classical properties, while in non-classical theories they have the striking peculiarities associated with quantum mechanics.

In chapter 7, an operator is added to the propositional logic of a theory so that some probability statements can be expressed in the resulting modal logic. For example, one can say in such a logic that a proposition is certain to be true after some measurement. The last section of this chapter scrutinises the arguments in favour of a 'quantum logic', rejecting their conclusions. One can agree that properties may be rather different in classical and quantum theories, and their modal logics are importantly different, but the propositional logic used by these theories is the same.

Finally, in chapter 8, we use this logical analysis to 'interpret' quantum mechanics. The paradoxes of chapter 1 are discussed again and analysed in a general, non-technical way. In section 2, the debate between Bohr and Einstein is briefly reviewed in the light of this analysis. This logical interpretation traces the quantum peculiarities to the structure of quantum theories, in particular to the weakness of the quantum state. The peculiarities show that our theory is weak, not that reality itself is bizarre.

Much of this discussion is rather technical. Where the terminology is particularly daunting in a section, a summary is provided at the end under the title 'Summing up'. This gives the gist of the discussion, for those unwilling to grapple with the full technical details, so that later discussion can be followed.

1 Classical and Quantum Mechanics

1.1 Mathematical details

In modern physics, two different systems of mechanics are used to describe the motion of physical things. Classical mechanics has been used since the time of Newton to describe the motion of ordinary bodies and of large systems like the rotating planets, but early this century it was found to be inaccurate when applied to very small systems such as the hydrogen atom, which has a radius of about 10^{-10} m. Quantum mechanics has since been developed to describe such microscopic systems.

Although classical and quantum mechanics have something in common—after all, they coincide when describing macroscopic reality—there are also fundamental differences between them. To understand these differences, and the problems they raise for our understanding of quantum mechanics, one must compare the mathematical structure of classical and quantum theories, so a brief discussion of this now follows. In §1.2 some of the problems and quantum 'paradoxes' are introduced in a less technical way.

1.1.1 Classical theories

Classical and quantum mechanics are alike in the following respects. Firstly, they describe reality in similar terms; for example both use the basic 'observables', position, momentum and energy in their descriptions. In both theories these observables take numerical values, which we assume can be obtained by measurement. In both cases observables are interrelated by specified laws, such as Newton's second law in classical mechanics, which asserts that 'force = mass × acceleration'. Observables take values which are consistent with these laws. As well as using similar observables, both classical and quantum theories describe a system at any time by assigning to it a state, though later we shall see that classical and quantum states have important differences. Also, both theories use an 'evolution equation', known in both cases as the Hamiltonian, which describes the continuous evolution of the system through time. Given the initial state of any system at time t_0, one can use this equation to compute the unique later states of the system at times t_1, t_2, provided that the system is left undisturbed. It is in this sense that both classical and quantum mechanics can be called deterministic theories.

However, there are important differences between classical and quantum mechanics. To appreciate their importance fully one must understand the different ways in which classical and quantum theories are expressed. We now very briefly review the mathematical representation of classical theories, before giving a more detailed account of the structure of quantum mechanics in the next section.

First, consider the simple case of a classical theory describing a system with just one degree of freedom. In this simple case the theory uses two fundamental observables, position and momentum, to describe a system at any particular time. Each observable is associated with a continuous set of values, $\{q_i\}$ and $\{p_j\}$ respectively. Any other observables can be defined in terms of these, by means of the classical laws. A state of the theory is determined when every observable is assigned a value, and so in this simple case, states correspond to ordered pairs of position and momentum values. The state of a system at time t is represented by an ordered pair of position and momentum values $(p(t), q(t))$. A system described by state (q', p') is said to have position q' and momentum p', and may have other observable properties derived from these.

In general, a classical theory can be represented mathematically in terms of a 'phase space', the space which has generalised position and momentum coordinates as parameters. A system with n degrees of freedom needs a phase space of $2n$ dimensions—n axes to represent the position coordinates $q_1, \ldots, q_n$, and n axes for the corresponding momentum coordinates $p_1, \ldots, p_n$. Classical observables are represented by real-valued functions of the phase variables. For example, the observable 'position in the direction i' for $1 \leqslant i \leqslant n$, is represented by the function Q_i, defined by the condition that $Q_i(q_1, \ldots, q_n, p_1, \ldots, p_n) = q_i$, and similarly for other position and momentum observables. Other classical observables, such as mass or energy, can be defined in terms of these.

Each state of a classical theory is represented by a point in phase space. A point $(q'_1, \ldots, q'_n, p'_1, \ldots, p'_n)$ assigns particular values q'_i, p'_j to the position and momentum observables Q_i, P_j, and hence assigns a value to every observable of the theory. The evolution of a system is described by an appropriate Hamiltonian H which is a function $H(q_1, \ldots, q_n, p_1, \ldots, p_n)$ of the phase variables. H represents the classical observable energy, and determines a path in the phase space for any initial state, by solving Hamilton's equations

$$\frac{\mathrm{d}p}{\mathrm{d}t} = -\frac{\partial H}{\partial q} \quad \text{and} \quad \frac{\mathrm{d}q}{\mathrm{d}t} = \frac{\partial H}{\partial p}.$$

When parametrised by time, the path represents successive states of the evolving system.

Classical states are idealised in the sense that they specify an exact value for every observable. In practice of course these may never be known precisely, since such measurements do not provide exact outcomes. However, the physical properties of actual measuring procedures are not relevant to the structure of classical mechanics. We always assume that a real system corresponds to some precise state, even if this can never be discovered. In fact, our ignorance about the state of a system can be expressed using a 'probability state', a probability assignment to the points in phase space. If probability p is assigned to a point in phase space, we understand p as a measure of the 'likelihood' that the system is in the state represented by this point. The classical probabilities will be discussed later in §§5.1 and 5.2.

1.1.2 Quantum theories

This classical use of probabilities to express ignorance about a state highlights the conceptual difficulties of quantum mechanics, for it is precisely this understanding which is not possible in quantum theories. Quantum states cannot be represented as points in a phase space, and quantum probabilities cannot be understood in the classical way as measures over this space. It is hard to see what quantum probabilities mean.

Quantum mechanics was first developed in the two separate forms of matrix mechanics and wave mechanics. Heisenberg developed his matrix mechanics in 1925, and Schrödinger soon followed with wave mechanics in 1926. It was soon proved by Schrödinger that these very different methods in fact gave the same descriptions of microscopic systems (see e.g. Jammer 1974). By 1932, von Neumann could demonstrate that these were essentially the same theory, and since then the term 'quantum mechanics' has been used to refer to both. Von Neumann later showed that matrix and wave mechanics are identical in the sense that both use different manifestations of abstract Hilbert Space (von Neumann 1955; see also Mackey 1963 ch 1, Bub 1974 ch 1).

Consider a physical system which would classically be described in terms of phase variables q and p, and which would be characterised by the classical Hamiltonian $H(q, p)$. According to the methods of matrix mechanics, this system is properly described not by the variables q, p, but instead by corresponding matrices, or arrays of complex numbers $\mathbf{Q}, \mathbf{P}$. Instead of the classical Hamiltonian $H(q, p)$, the energy of the system is described by the matrix $\mathbf{H}(\mathbf{Q}, \mathbf{P})$. To discover the allowable 'energy levels', or values of $\mathbf{H}$, one must find matrices $\mathbf{Q}, \mathbf{P}$, satisfying the commutation relations

$$\mathbf{PQ} - \mathbf{QP} = i\hbar$$

for which the Hamiltonian matrix $\mathbf{H}(\mathbf{Q},\mathbf{P})$ is diagonal. This involves finding the orthogonal matrix $\mathbf{S}$ such that

$$\mathbf{S}^{-1}\mathbf{H}\mathbf{S} = \mathbf{E}$$

where $\mathbf{E}$ is $\mathbf{H}$ in diagonal form. The diagonal elements of $\mathbf{E}$, the eigenvalues, give the allowable energy levels of the system. The corresponding eigenvectors are columns $(s_{11}, s_{21}, \ldots)$ $(s_{12}, s_{22}, \ldots)$ of the matrix $\mathbf{S}$, from which may be deduced further properties of the system.

According to wave mechanics, on the other hand, the same system would be described using complex-valued functions $\psi(q)$ of the position variable. Corresponding to the classical phase variables q, p, wave mechanics uses operators Q, $-i\hbar\,\partial/\partial Q$ which act on these wavefunctions. The classical Hamiltonian $H(q, p)$ is replaced by a corresponding functional operator $H(Q, -i\hbar\partial/\partial Q)$. The allowable energy levels of the system are found by solving the equation

$$H(Q, -i\hbar\,\partial/\partial Q)\psi(q) = e\psi(q).$$

The eigenvalues e give the allowable energy levels of the system, while the wavefunctions $\psi(q)$ are corresponding eigenvectors which may be used to represent the state of the system.

Von Neumann showed that the two theories are identical, by showing that the space l^2 of sequences $(s_{11}, s_{21}, \ldots)$, $(s_{12}, s_{22}, \ldots)$ used in matrix mechanics, is essentially the same as the space L^2 of wavefunctions $\psi(q)$. The two spaces are isomorphic[†]. Operators Q, P, on l^2 correspond to operators Q, $-i\hbar\,\partial/\partial Q$ on L^2, and these operators satisfy the same commutation relations. The Hamiltonian matrix $\mathbf{H}(\mathbf{Q},\mathbf{P})$ on l^2 corresponds to the operator $H(Q, -i\hbar\,\partial/\partial Q)$ on L^2. Similarly, any operator which represents an observable on one space will correspond under the isomorphism to an operator representing the same observable on the other. Although they seem different, the theories are exactly equivalent.

1.1.3 *The Hilbert Space representation*

The German mathematician David Hilbert (1862–1943) had already studied the spaces l^2 and L^2, and so the term Hilbert Space is used for all such structures. By showing how quantum mechanics could be formulated in terms of abstract Hilbert Space, von Neumann displayed the essential features of this theory. A brief review of this formulation is now given, so that the structure of quantum mechanics can be compared with that of a classical theory. First, some definitions for the analysis of abstract Hilbert Space are given, and then these will be related to the discussion of

---

[†] An isomorphism is a one-to-one, structure preserving map. For details of the precise nature of this isomorphism, see von Neumann (1955 ch 1). This is also discussed more briefly by Bub (1974 ch 1).

quantum mechanics. It is important to remember that this abstract formulation of the theory refers equally to wave or matrix mechanics, by means of the correspondences above.

A *Hilbert Space* $\mathcal{H}$ is a linear vector space over the field of complex numbers $\mathbb{C}$, on which an inner product is defined, and which is complete with respect to the metric generated by this inner product. Since it is a *vector space*, elements of $\mathcal{H}$ are vectors ψ, ϕ, which can be added together, or multiplied by complex numbers (scalars). Since $\mathcal{H}$ is *linear*, it is closed with respect to these operations, so if ψ, ϕ, are in $\mathcal{H}$, $\psi + \phi$ is also a vector in $\mathcal{H}$, and so too is $c\psi$ for any complex number c. The *inner product* is a map taking ordered pairs of vectors into the complex numbers, so that any inner product (ψ, ϕ) is a number in $\mathbb{C}$. This map meets the conditions

$$(\psi, \phi) = \overline{(\psi, \phi)}, \text{ its complex conjugate}$$
$$(\psi, \alpha + \beta) = (\psi, \alpha) + (\psi, \beta)$$
$$(\psi, c\phi) = c(\psi, \phi)$$
$$(\psi, \psi) \neq 0, \text{ unless } \psi = 0, \text{ the null vector.}$$

Two vectors are *orthogonal* if their inner product is zero. The *norm* (or magnitude) of a vector ψ is $\|\psi\|$, where $\|\psi\|^2 = |(\psi, \psi)|$, the absolute value of the inner product of ψ with itself. ψ is a *normed unit* vector if $(\psi, \psi) = 1$. A metric can be defined on any Hilbert Space by associating with any pair of vectors ψ, ϕ the norm of their difference, $\|\psi - \phi\|$. The properties of the inner product ensure that this is indeed a metric, i.e. a distance function, in the usual sense. For example, one can easily show that $\|\psi - \phi\| = 0$, if and only if $\psi = \phi$, so there is zero distance between vectors only when they are identical. A Hilbert Space is *complete* with respect to this metric i.e. it must contain all its limit points[†].

According to von Neumann's formulation, the 'pure' states of a quantum theory are represented by the normed unit vectors of some Hilbert Space associated with the theory. (The representation is not unique, in that two normalised vectors, differing by a complex multiple, give the same state.) States which are not pure, the quantum 'mixtures', are weighted sums of pure states. In fact, a Hilbert Space can represent a theory only if it is also separable (i.e. has countable dimension, see below). Similarly, operators in the space must have special properties if they are to represent quantum observables. These 'observable operators' must be linear and self-adjoint. The exact nature of these properties will not concern us here (see von Neumann (1955) and also Halmos (1951), Bub (1974) and Gillespie (1973)).

As in classical theories, quantum observables are associated with a set

---

† In other words, every sequence that is Cauchy convergent converges to a point in the Hilbert Space.

of real numbers which are their possible measured values. This set is the spectrum of the operator representing that quantum observable. In the simplest case this means that the allowed values of observable A, resented by the operator $\hat{A}$, are the solutions a to the eigenvalue equation for $\hat{A}$:

$$\hat{A}\alpha_i = a\alpha_i$$

for some eigenvector α_i of $\hat{A}$. The self-adjoint property of observable operators ensures that their eigenvalues are always real rather than complex numbers and so, as one would expect, the result of measuring an observable will always be a real number.

It is a peculiarity of quantum theories that observables may have discrete rather than continuously distributed values. Indeed, it was the discovery of discrete energy levels in some atomic systems which gave quantum mechanics its name. This peculiarity is evident in the mathematical structure of the theory, since there are observable operators in Hilbert Space which have discretely distributed eigenvalues. In general the spectrum of an observable operator may be continuous or it may be discrete or it may be made up of discrete and continuous portions.

To understand how a vector in Hilbert Space can represent a quantum state, one must understand how vectors can be decomposed into 'components' along axes in the space. Each observable operator determines a set of axes in the Hilbert Space representing a quantum theory†. The probability assigned by any quantum state to each value of that observable will depend on the magnitude of the component of the state vector, along the appropriate axis in the space. To express this precisely some new terms must be introduced.

A set $\{\alpha_i\}$ of vectors in $\mathcal{H}$ are *orthogonal* if every pair of vectors in the set is orthogonal, that is if $(\alpha_i, \alpha_j) = 0$, whenever $i \neq j$. This set is *orthonormal* if it is both orthogonal and also every vector has unit norm, that is $|(\alpha_i, \alpha_i)| = 1$ for each i. The set is *complete* if any vector in $\mathcal{H}$ can be expressed as a linear combination of these vectors. That is, there must be a set $\{c_i\}$ of scalars such that

$$\psi = \sum_i c_i \alpha_i$$

for any vector ψ in $\mathcal{H}$. A set $\{\varepsilon_i\}$ of vectors in $\mathcal{H}$ is a basis (or set of axes) if it is both orthonormal and complete. The *dimension* of a space is given by the number of vectors in any basis set. (The dimension is unique since

† The operator defines a unique set of axes only if the spectrum is simple i.e. only if it contains no degenerate eigenvalues. The case where an eigenvalue is degenerate is discussed in §6.3.3.

every basis set has the same number of elements.) It can be shown that, given any basis of $\mathcal{H}$, any vector can be expressed

$$\psi = \sum_i (\varepsilon_i, \psi)\varepsilon_i.$$

The vectors $(\varepsilon_i, \psi)\varepsilon_i$ are the *components* of ψ in the basis $\{\varepsilon_i\}$. The scalars (ε_i, ψ) thus give the magnitudes of these components. For example $|(\varepsilon_i, \psi)|$ is the magnitude of the ith component of ψ in the basis $\{\varepsilon_i\}$. Any vector can be decomposed in this way, into a sum of component vectors along the axes of any basis set.

If an observable operator $\hat{A}$ on Hilbert Space $\mathcal{H}$, has discrete eigenvalues $\{a_i\}$, then it will have a complete orthonormal set of eigenvectors $\{\alpha_i\}$. So this set of eigenvectors forms a basis for $\mathcal{H}$, called the *eigenbasis* of operator $\hat{A}$. The a_i are its eigenvalues and the α_i its eigenvectors. Since the eigenvectors form a basis, any vector can be decomposed into components along these axes thus:

$$\psi = \sum_i (\alpha_i, \psi)\alpha_i.$$

It is this decomposition which generates the probabilities assigned by the quantum state ψ to the values of observable A. Each value a_i of observable A is an eigenvalue of operator $\hat{A}$, and so has an axis (that is an eigenvector) in the eigenbasis of $\hat{A}$ to represent it. The probability, according to state ψ, that a measurement of A will yield result a_i is given by

$$prob_\psi\,(a_i:A) = |(\alpha_i, \psi)|^2.$$

This is the square of the magnitude of the component of ψ in the direction α_i. This expression for the quantum probability is Born's interpretation of the state vector. According to Born's interpretation, this probability assignment is in fact the strongest assertion made by state ψ about value a_i of A.

Note that in the case where operator $\hat{A}$ has discrete eigenvalues which are not always distinct (i.e. $\hat{A}$ has degenerate eigenvalues), then the eigenbasis is not unique. For example, suppose just two eigenvalues coincide, so that $a_i = a_j$, for $i \neq j$, and $a_i \neq a_k$ for any other k. Then in this case value a_i is associated with a two-dimensional subspace, and any pair of orthogonal vectors in this subspace will serve as vectors in the eigenbasis of $\hat{A}$. It follows that the probabilities associated with the degenerate eigenvalues will depend on the choice of vectors in the basis set. If an observable operator $\hat{A}$ has eigenvalues which are not always discrete (i.e. A has a spectrum which is at least partly continuous), then a rather more complex analysis is needed than that given above. In fact, one reformulates the expressions above in terms of projection operators, and this formulation can be extended to the continuous case. Hence an analogous expression for probabilities will apply, even when the spectrum of $\hat{A}$ is continuous. Although the discrete case already described is enough to show the conceptual interest of quantum mechanics, the reformulation in terms of projection operators will be useful later, and so is briefly described.

Firstly, one must introduce some new terms. A subset $\mathscr{S}$ of vectors in $\mathscr{H}$ is a *subspace* of $\mathscr{H}$ if $\mathscr{S}$ is itself a Hilbert Space. That is, $\mathscr{S}$ must be closed with respect to vector addition and scalar multiplication. Any vector in $\mathscr{H}$ is an example of a one-dimensional subspace. $\mathscr{S}^{\perp}$, the *orthogonal complement* of $\mathscr{S}$, is that subspace which contains all vectors orthogonal to all vectors in $\mathscr{S}$. Any vector in $\mathscr{H}$ can be uniquely decomposed into a sum of vectors from two orthogonal subspaces. That is, for any $\mathscr{S}$ and any ψ

$$\psi = \psi_{\mathscr{S}} + \psi_{\mathscr{S}^{\perp}}$$

for $\psi_{\mathscr{S}}$ in $\mathscr{S}$, and $\psi_{\mathscr{S}^{\perp}}$ in $\mathscr{S}^{\perp}$. The $\mathscr{S}$ component of ψ, $\psi_{\mathscr{S}}$, is the *projection of ψ in $\mathscr{S}$*. In general one can define a *projection operator* $\hat{P}_{\mathscr{S}}$ associated with any subspace $\mathscr{S}$, by setting

$$\hat{P}_{\mathscr{S}}\psi = \psi_{\mathscr{S}}$$

for any ψ. Projection operators have the special properties of observable operators, and in addition are idempotent, that is to say $\hat{P}^2 = \hat{P}$, for any projection operator $\hat{P}$. There is clearly a one-to-one correspondence between subspaces and projection operators. Each projection operator corresponds to the subspace which is its range.

The earlier decomposition of an observable operator $\hat{A}$ in terms of its eigenbasis can be re-expressed as the sum

$$\hat{A} = \sum_{i} a_{i}\hat{P}_{a_{i}}$$

where $\hat{P}_{a_{i}}$ is the projection operator corresponding to the subspace associated with eigenvalue a_{i}: this is just the subspace which consists of all vectors α satisfying the equation $\hat{A}\alpha = a_{i}\alpha$. (If $\hat{A}$ is simple, with n distinct eigenvalues $a_{1}, \ldots, a_{n}$, then $\hat{P}_{a_{i}}$ simply corresponds to the one-dimensional subspace which is the vector α_{i}.) The components of any vector ψ in the eigenbasis of $\hat{A}$ can now be expressed in terms of projections, since

$$(\alpha_{i}, \psi)\alpha_{i} = \hat{P}_{a_{i}}\psi.$$

This component is just the projection of ψ into the subspace associated with a_{i}. The probability assigned by the quantum state ψ to this value of A is given by the magnitude of this projection thus:

$$prob_\psi \, (a_i : A) = \| P_{a_i} \psi \|^2.$$

An operator $\hat{A}$ with continuous spectrum does not strictly speaking have eigenvectors, and for this reason one must generalise the expressions given in terms of projection operators. The generalisation is quite straightforward. According to the Spectral Theorem, any observable operator $\hat{A}$ in a Hilbert Space may be expressed in terms of the following decomposition

$$\hat{A} = \int_{-\infty}^{+\infty} \lambda \; dE_\lambda$$

where the spectral family $\{E_\lambda\}$ is a set of projection operators which must meet certain requirements analogous to the properties of a basis set, and which are indexed by the members λ of the spectrum of A. Note that in the case of an operator with distinct eigenvalues $\{a_i\}$, the projections E_λ correspond to the operators $\hat{P}_{a_i}$ which have the eigenvectors α_i as their range. The general expression for the probabilities assigned to ranges of values of A is

$$prob_\psi \, (\lambda_2 - \lambda_1 : A) = \| E_{\lambda_2} \psi - E_{\lambda_1} \psi \|^2.$$

This gives the probability according to the quantum state ψ, that a measurement of observable A will yield a value in the interval between λ_2 and λ_1. This expression will apply for any observable operator, including those with a continuous spectrum. For simplicity, we shall generally use the earlier discrete expressions to illustrate examples.

1.1.4 *Quantum states*

According to Born's interpretation, these probability assignments are the most the theory can say in state ψ about the system. For example, in quantum state ψ one can say no more about the value a_i of a simple observable A, than that $prob_\psi \, (a_i : A) = |(\alpha_i, \psi)|^2$. By now this is generally accepted as a fundamental principle of quantum mechanics, but it is a principle which raises important interpretative problems, for it follows from the mathematical structure outlined above, that quantum states are essentially statistical, in the sense that they can make only statistical predictions about the values of quantum observables. This is of course quite unlike the classical situation, where states assign precise values to all observables.

Of course, there are special situations where a quantum state does assign probability 1 to the value of an observable, and so the outcome of a measurement is predicted with certainty. This happens when a system is described by a state which is an eigenvector of the observable considered, in which case we say that the system is in an *eigenstate* of that observable, for we can see from the analysis above, that when a system is in, for example, state α_i, the eigenvector corresponding to value a_i of A, then in this state the probability of finding value a_i after a measurement of A will be 1. This is because

$$prob_{\alpha_i}(a_i : A) = |(\alpha_i, \alpha_i)|^2 = 1,$$

since α_i is a normed unit vector. It is a mistake, however, to think that such eigenstates are privileged, or 'closer' to classical states than other states of the quantum theory, for quantum mechanics does not itself discriminate, regarding all normed unit vectors alike as representing pure states of the theory. Secondly even these eigenstates are essentially statistical where other observables are concerned. Although state α_i is an eigenvector of $\hat{A}$, and so predicts the value a_i of this observable with certainty, α_i generally fails to be also an eigenstate of some other observable operator $\hat{B}$, and so will make only statistical predictions about the values of this observable.

If two observables do have members of their eigenbasis in common, then it will sometimes be possible to make predictions with certainty about values of both observables. For example, if ψ_i is an eigenstate both of $\hat{A}$ and of $\hat{B}$, so that ψ_i corresponds to eigenvalue a_i of A and b_i of B, then using the earlier argument

$$prob_{\psi_i}(a_i : A) = prob_{\psi_i}(b_i : B) = 1$$

and so in this quantum state both the values a_i and b_i are predicted with certainty. If two operators have an entire eigenbasis in common, then the observables they represent are said to be *compatible* (see §4.2.4). In this case predictions with certainty can be made about values of both observables at once, in any pure quantum state. That is, when A and B are compatible, any pure state which assigns probability 1 to a value of A will also assign probability 1 to some value of B. Because of this property, compatible observables are sometimes referred to as *simultaneously measurable*. Equivalently, one can define compatibility not in terms of a shared eigenbasis, but in terms of the commutation of observable operators. $\hat{A}$ and $\hat{B}$ are said to *commute* if

$$\hat{A}\hat{B}\psi = \hat{B}\hat{A}\psi \quad \text{for any vector } \psi.$$

It turns out that observables are compatible in the earlier sense if, and only if, their operators commute.

It is a peculiarity of quantum mechanics that it uses observables which are not compatible. Thus states do not generally assign probability 1 to some value of every observable. Position and momentum are examples of incompatible quantum observables, since the operators $\hat{Q}$ and $\hat{P}$ representing them do not share an eigenbasis or, equivalently, do not commute. So a state which assigns probability 1 to some position value will not do so to any momentum value, and vice versa. Only statistical predictions can be made about these values. In fact, Heisenberg's

Uncertainty Principle specifies the mathematical relation holding between the probabilities assigned to incompatible quantum observables. The operators $\hat{Q}$ and $\hat{P}$ which represent position and momentum satisfy the following commutation relation

$$\hat{P}\hat{Q} - \hat{Q}\hat{P} = i\hbar\hat{I}$$

where $\hat{I}$ is the unit operator. It can be proved as a theorem about observable operators that if $\hat{A}\hat{B} - \hat{B}\hat{A} = a\hat{I}$, then the product of the uncertainties $\Delta\hat{A} \, \Delta\hat{B} > a/2$. (Here $\Delta\hat{A}$ is the dispersion of A, the standard deviation of the expectation value of the observable. A formal definition is given in §5.1.2.) It follows, in the case of position and momentum operators, that

$$\Delta\hat{Q}\,\Delta\hat{P} \geqslant \hbar/2.$$

This is Heisenberg's uncertainty relation for position and momentum. The smaller the uncertainty associated with one observable, the larger that associated with the other. There is no state in which both are dispersion-free.

Thus we see that quantum states differ from the states of classical mechanics. A classical state corresponds to a point in classical phase space, and hence to a precise assignment of values to all observables. On the other hand a pure quantum state corresponds to a unit vector in Hilbert Space, and hence, as we have seen, to an assignment of probabilities to the values of observables. The existence of incompatible observables in quantum theories means that quantum states are irreducibly statistical. They do not generally assign probability 1 to some value of every observable.

It follows that there is a problem in understanding quantum probabilities which does not arise in classical theories. Classical probabilities may be represented as measures over classical phase space, and can be understood as expressions of ignorance about the state of a system. But in quantum theories, as we shall discuss in more detail below, there is seemingly no analogue of classical phase space. Since quantum states are themselves statistical, probabilities do not simply express ignorance of the state of a system. If quantum probabilities do express ignorance, we are faced with the further question, 'Ignorance of what?'

Quite apart from these problems, there is another feature of quantum theories which raises fundamental problems of interpretation. According to quantum mechanics, the state of any system changes abruptly but predictably after a measurement is performed. The rule describing this change is known as von Neumann's Projection Postulate, and is generally accepted as a principle of quantum mechanics. According to the Projection Postulate, the state of a system changes after measurement in such a way that the measurement outcome, which may not have been certain beforehand, is now certain in the new state used after the measurement. To quote von Neumann himself, who takes $\mathscr{R}$ as a magnitude (von Neumann 1955 p 214):

> ... if the system is initially found in a state in which the values of $\mathscr{R}$ cannot be predicted with certainty, then this state is transformed by a measurement M of $\mathscr{R}$ ... into another state: namely, into one in which the value of $\mathscr{R}$ is uniquely determined. Moreover the new state, in which M places the system, depends not only on the arrangement of M (which could not be predicted causally in the original state)—because the value of M in the new state must be actually equal to this result.

Consider for example the simple case of a magnitude A which has n distinct discrete eigenvalues, $a_1, \ldots, a_n$, and $\psi = \Sigma_i a_i \alpha_i$. Suppose a measurement M of A on a system initially in state ψ yields outcome a_i. According to the Projection Postulate above, after M the state of the system must be such that the value a_i is predicted with certainty. According to our earlier discussion this new state must be the eigenstate of $\hat{A}$ corresponding to eigenvalue a_i, so in this case the new state of the system is α_i, since it is in state α_i that $prob_{\alpha_i}(a_i:A) = 1$. Since the initial state ψ may be quite different from α_i, the Projection Postulate generally describes a drastic alteration in the state of the system. The quantum state appears to 'jump' after a measurement is performed.

Like classical mechanics, quantum theory does incorporate an 'evolution equation', or 'equation of motion' which describes the continuous change of any state through time. In quantum mechanics, the Hamiltonian operator H, representing the energy observable, determines the time evolution of any state through the differential equation

$$\hat{H} = i\hbar \frac{\partial \psi}{\partial t}.$$

Given any initial state ψ one can compute by means of this equation the successive states $\psi_1, \psi_2, \ldots$ of the same system at successive times $t_1, t_2, \ldots$, provided that the system is left undisturbed. However, this equation will not account for the discontinuous changes of state described by the Projection Postulate.

Thus quantum mechanics differs from classical mechanics, not only in that states are essentially statistical, but also because the state of a system changes in a drastic but well-defined way after a measurement takes place. Furthermore, the alteration is such that the value of an observable is known with certainty only in the altered state, after the measurement is performed. So in general the value of a quantum observable is known only after measurement, and is not assumed to have the value in the state preceding measurement. This is quite unlike the situation in classical mechanics, where states are not supposed to alter after measurement, indeed where measurements are assumed to 'discover' the pre-measured state of a system.

1.1.5 Summing up

We have seen that classical theories permit a 'phase space' representation in which every state is associated with some point, and hence with a precise value for each observable. However, in quantum theories there appears to be no phase space. Instead states are represented by vectors in Hilbert Space, and observables by operators on the vectors. This representation was discussed in some detail.

Two fundamental non-classical properties of quantum states were discussed. The first was their essentially statistical nature and the existence of incompatible quantum observables. The second was the 'jump' which occurs in the state of a system after a measurement is performed. These peculiarities lie at the heart of some well-known quantum 'paradoxes' which will now be discussed.

1.2 Conceptual problems and some quantum paradoxes

1.2.1 Early disturbance interpretations

The simple 'disturbance' interpretation of quantum mechanics attributes all oddities of the theory to disturbances caused by measurement. Heisenberg and Bohr, i.e. the 'early Copenhagen' school, were the early advocates of this view, and it is still popular today. They argued that the incompatibility of some observables in quantum theories reflects the fact that their simultaneous measurement is physically impossible. Thus, for example, position and momentum are incompatible in quantum mechanics because any measurement of one in a microscopic system destroys evidence of the other. A detector set in the path of an electron will register its position only at the expense of deflecting it, in which case the original momentum cannot be measured.

According to the 'naive' disturbance interpretation, Heisenberg's Uncertainty Principle expresses just this sort of physical impossibility. Any measurement of one observable leaves us in ignorance of the other, so that it is physically impossible to be certain about all properties at once. The 'jump' in the state of a system which may occur after measurement is also, on this view, a consequence of the physical disturbance caused by measuring instruments. The change in state is supposed to describe a physical alteration caused by measurement.

This version of the disturbance interpretation is attractively simple. However, to some extent it obscures rather than solves the main problem raised by quantum theory. For instance, in classical mechanics practical problems of measurement are irrelevant to the concept of a classical state, since this is defined abstractly in terms of the values of observables. Why is there no analogous concept of state in quantum mechanics? Secondly, this simple disturbance interpretation does not make clear what quantum theories tell us about unmeasured systems, and hence we still do not know how quantum probabilities are understood. Suppose for example, that quantum state ψ assigns non-zero probabilities both to the values of observable A and also to values of an incompatible observable B. One can agree with the disturbance interpretation that no measurement procedure can actually determine the correct value of A and also that of B, but what does ψ tell us about the system if it is not measured?

To answer this kind of query the disturbance interpretation is usually elaborated. For example it is sometimes claimed that measurements not only disturb microscopic systems, they also in some sense 'create' the property which is measured. According to this account, quantum systems, unlike their classical counterparts, have in fact no properties at all until they are measured. An unmeasured system is supposed to have only 'dispositions' which become 'actualised' by measurement procedures. Thus the probabilities assigned by ψ tell us how likely certain properties are to be actualised by appropriate measurements.

1.2.2 Schrödinger's cat

The paradox known as 'Schrödinger's cat', shows up the difficulties of even such a modified disturbance interpretation of quantum theories[†]. In this example Schrödinger considers a microscopic experiment which is coupled to a macroscopic system. In fact one can consider a cat (the macroscopic system) inside a glass box, in the corner of which is a 'hellish contraption' (the microscopic system), consisting of an electron source in front of a half-silvered mirror, which is placed before a tiny hammer suspended over a small vial of cyanide. This contraption is designed so that when any electron passes through the mirror it hits the hammer, which falls on the vial, releasing enough cyanide into the box to kill the unfortunate cat. The experiment is straightforward so long as a beam of electrons is fired at the mirror, since in this case half the electrons are transmitted through it, so the hammer falls and the cat is killed. Schrödinger's paradox arises if the beam of electrons is supposed reduced until electrons are emitted just one at a time. If a single electron is fired at the mirror, how do we describe the system after some elapse of time?

Classically the experiment is clear. There is a probability $\frac{1}{2}$ that the electron is reflected from the mirror, and $\frac{1}{2}$ that it is transmitted. So

---

† References to Schrödinger's original paper and other discussions of this experiment can be found in Jammer (1974) pp 217ff.

according to classical mechanics the electron is either in the reflected state, say S_R, or it is in the transmitted state S_T after the experiment. So long as no measurement is made, each state is assigned probability $\frac{1}{2}$, expressing our ignorance of the actual state of affairs. Once the system is observed, however, then either S_R or S_T will be recognised as the true state of the electron. Thus we will know either that the cat lives, or that it is dead, and that it was already that way before observation.

Quantum mechanics gives a different account of this experiment. Let ψ_R and ψ_T be the two quantum states analogous to S_R and S_T, corresponding to the reflection and transmission of the electron respectively. (ψ_R, ψ_T will be eigenstates of an appropriate position operator.) After the experiment is performed, before any measurement takes place, it seems that the state of the electron is the linear combination or 'superposition' of states ψ', where

$$\psi' = \sqrt{1/2}\,\psi_R + \sqrt{1/2}\,\psi_T.$$

This state is quite different from either ψ_R or ψ_T. One can, however, deduce from ψ' that after measurement it will 'jump' to one of these two states, with probability $\frac{1}{2}$ assigned to each. So, contrary to classical theory, quantum mechanics insists that after the experiment is performed and before any measurement is made, the electron is neither in state ψ_R nor in state ψ_T, but is instead in the superposition ψ'.

Schrödinger's paradox arises when one considers the fate of the cat after the experiment. The state of the cat clearly depends on that of the electron and so it seems that the cat too must be in a superposition ϕ', where

$$\phi' = \sqrt{1/2}\,\phi_A + \sqrt{1/2}\,\phi_D$$

and ϕ_A, ϕ_D are the states of being alive and being dead respectively. It seems that the cat is neither dead nor alive after the experiment, and this of course contradicts our everyday knowledge of cats. We might think we know what it means for an electron to be in a superposition of states, but this notion is plainly paradoxical when applied to cats. On the other hand, if we accept that some time after the experiment the cat must be either quite dead or really alive, then quantum theory is contradicted, for in this case one could deduce that the electron is correspondingly either in state ψ_R or in ψ_T, and this is contrary to quantum theory, which finds it in state ψ'.

The second aspect of this paradox concerns the nature of measurement in quantum systems. After a measurement is performed on the system, the state of the electron 'jumps' from the superposition ψ', to either ψ_R or ψ_T with equal probability for each occurrence, and presumably a similar 'jump' occurs in the state of the cat. However, in this example a measurement may be simply a glance at the glass box. How can a glance at the cat disturb it to such an extent that it alters entirely from being in a superposition, to being either quite dead or decidedly alive? Perhaps even more peculiar, how can a glance at the cat disturb the electron, by now perhaps far away, so that it too alters its state? The naive disturbance interpretation seems, in this case, a little too naive.

It is sometimes argued that subsystems are never described by a pure quantum state. In this case it is claimed that only the entire system of cat, box and 'hellish contraption' is properly described by a superposition of states. On this view this particular paradox is resolved, in the sense that it is ruled illegal. The cat is not individually described by the theory and so no paradox can arise from its description. However, it is clear that a wider paradox does remain. We still do not understand how the entire system differs from a straightforward macroscopic system. Why is it described by a superposition, and what does this description mean? What for example, does a superposition tell us about subsystems, such as the cat in this example? And what does the 'jump' in the state of the system signify?

By linking a microscopic system with a macroscopic system in his example, Schrödinger attempts to highlight the conceptual difficulties of quantum descriptions. If we are to accept quantum mechanics completely, then we have to accept that microscopic 'objects' such as electrons are quite unlike our everyday objects such as cats.

Indeed this is one 'hard-nosed' reaction to the paradox which is sometimes proposed as a solution: cats are not proper objects of quantum theory. They are 'classical' elements of macroscopic reality which are not properly described by quantum states at all. It is supposed to be our improper attempt to do this which generates Schrödinger's paradox.

According to this view also, the cat is not described after the experiment by a pure state as declared earlier. It is now supposed to be described by a 'mixture' expressing classical ignorance about its condition. Sometimes this is supported by a variation of the 'disturbance' interpretation, with an argument that the apparatus involved is a measuring instrument, and as such can yield only 'classical' descriptions. On this view, one presumably accepts that a glance at the cat does not disturb it and change its state. Instead, the whole apparatus itself is supposed in some way to 'disturb' the system and yield only classical descriptions. We shall see in later discussions of Bohr, and of the EPR experiment, that this view has problems of its own. It is also rather unhelpful. It seems that Schrödinger's paradox is 'resolved' by deciding that cats are fundamentally different from electrons, and so cannot be described by subatomic theory, but it seems also that this difference cannot be measured or even understood. We must apparently accept that microscopic reality is completely unknown to us. The difficulty with this 'solution' is that we still do not understand the experiment. In a sense we have simply been told not to try.

1.2.3 The EPR experiment

Throughout his life Einstein remained completely opposed to the kind of view which accepts the paradoxes as marking the limits of our understanding. He argued instead that examples such as Schrödinger's reveal that quantum theories are inadequate, in the sense that their descriptions are incomplete. According to his view, the cat is indeed either dead or alive after the experiment, it is just that quantum mechanics fails to tell us which.

In order to support his own view, Einstein proposed another kind of 'thought experiment' which has since become widely known as the Einstein–Podolsky–Rosen, or EPR paradox. Einstein first published a discussion of his example in a paper written with these collaborators, entitled *Can Quantum Mechanical Description of Reality be Considered Complete?* (Einstein *et al* 1935). However, two years before this paper appeared, he had described a simple version of the paradox in conversation:

> What would you say of the following situation? Suppose two particles are set in motion towards each other with the same, very large, momentum, and that they interact with each other for a very short time when they pass at known positions. Consider now an observer who gets hold of one of the particles, far away from the region of interaction, and measures its momentum: then, from the condition of the experiment, he will obviously be able to deduce the momentum of the other particle. If, however, he chooses to measure the position of the first particle, he will be able to tell where the other particle is. This is a perfectly correct and straightforward deduction from the principles of quantum mechanics: but is it not very paradoxical? How can the final state of the second particle be influenced by a measurement performed on the first, after all physical interaction has ceased between them?

Here are the essential features of the EPR paradox: two microscopic systems are allowed to interact for a time, and then separate. The interaction establishes correlations between observables on the two systems, so that for example a measurement of the position of one determines also the position of the other, separated system. It follows, according to the Projection Postulate, that a measurement on one system causes the state of both to 'jump'. This seems paradoxical, since the systems are by now separate. And, perhaps even stranger, the choice of measurement performed on one determines to some extent the kind of 'jump' occurring in the states of both systems, hence also in the far distant one. For example, if position is measured, then the state of the other system also must be an eigenstate of this observable, while if we choose to measure momentum the value of this observable too will be determined on the other system. Thus the example raises the following problems for any disturbance interpretation: how can a measurement on one system drastically alter the state of another separated system? And how can the choice of measurement performed influence the kind of alteration which takes place?

There is a simpler experiment to illustrate the paradox than that described in the EPR paper (Bohm and Aharonov 1957). This concerns a molecule of total spin zero, consisting of two atoms each with spin $\frac{1}{2}$. The two atoms interact initially and are collectively in the state which is the superposition

$$\sigma = \sqrt{1/2}\psi_{\theta+}\phi_{\theta-} - \sqrt{1/2}\psi_{\theta-}\phi_{\theta+}$$

for states $\psi_{\theta+}, \psi_{\theta-}$, corresponding to atom 1 having spin $+\frac{1}{2}, -\frac{1}{2}$ respectively in direction θ, and states $\phi_{\theta+}, \phi_{\theta-}$ corresponding to similar properties of atom 2. For example, if we consider the direction x then we know from σ that either atom 1 has spin $+\frac{1}{2}$ and atom 2 spin $-\frac{1}{2}$ in this direction, or else atom 1 has spin $-\frac{1}{2}$ and atom 2 spin $+\frac{1}{2}$ in direction x, since

$$\sigma = \sqrt{1/2}\psi_{x+}\phi_{x-} - \sqrt{1/2}\psi_{x-}\phi_{x+}.$$

The total spin in any direction is always zero.

When the two atoms have separated, we suppose a measurement of spin in some direction is made on one of the atoms. According to quantum mechanics, after a measurement the state will 'jump' to one of the terms in the superposition. For example, if we measure the spin of atom 1 in direction x, and find that spin in this direction is $+\frac{1}{2}$, then we known from the projection postulate that σ changes to the state $\sigma' = \psi_{x+}\phi_{x-}$ after the measurement. It seems then that atom 1 is now in state ψ_{x+}, and that atom 2 is in state ϕ_{x-}. So here we encounter the first part of the paradox. How can the measurement of spin on atom 1 disturb the separated atom 2 to such an extent that its state 'jumps' from a superposition to state ϕ_{x-}?

The second part of the paradox arises because the spin components of an atom, say in the x, y and z directions, are incompatible. Thus the choice of measuring spin in the x direction on atom 1 determines that it is only the spin in this direction which can be known with certainty for either atom, for after this measurement the state of atom 2 also can only 'jump' either to ϕ_{x+} or else to ϕ_{x-}. Furthermore, after a measurement σ ceases to be the state of the system, and so the correlations which σ expresses no longer apply. Thus the choice of measurement performed on atom 1 not only determines which direction of spin has a well-defined value for atom 2, it also destroys the correlations between other spin components. How can our seemingly free choice of measurement on one atom determine so radically the kind of properties found on the other?

The EPR experiment clearly shows that a classical 'ignorance' or 'ensemble' understanding of the quantum probabilities cannot apply. One cannot, for example, say that before measurement 'atom 1 has spin $+\frac{1}{2}$ in the x direction, or spin $-\frac{1}{2}$ in this direction, but we don't know which', for we may in fact choose to measure spin in some other direction, say y, in which case neither disjunct in the statement about the x component of spin can be true, for the spin components in different directions are incompatible. This experiment also clearly raises difficulties for any straightforward disturbance interpretation of quantum probabilities. We cannot say that the measurement of spin in some direction physically alters the atoms so that their states 'jump' from the superposition to one of its components. Nor can it be simply a physical disturbance which destroys the correlations between other spin components, for the separated atom does not in fact interact with the measuring apparatus.

Einstein used the EPR paradox to argue in the following way that quantum descriptions of reality are incomplete. He adopted the following definitions as premises of the argument. First he defines what it is for a theory to be *complete* (Einstein *et al* 1935)[†]:

> The following requirement for a complete theory seems to be a necessary one: every element of the physical reality must have a counterpart in the physical theory.

Next he defines an *element of reality*:

> If without in any way disturbing a system we can predict with certainty (i.e. with probability equal to unity) the value of a physical quantity, then there exists an element of physical reality corresponding to this physical quantity.

Thus observables represent elements of reality if they are assigned probability 1 in some state of a quantum theory. Lastly, Einstein assumed what has since become known as the *Principle of Locality*:

> [If] at the time of measurement . . . two systems no longer interact, no real change can take place in the second system, in consequence of anything that may be done in the first.

That is, there can be no action at a distance.

From these premises, and the EPR experiment, Einstein argues that

---

[†]   These excerpts are reprinted by permission of the Hebrew University of Jerusalem.

quantum mechanics is not a complete theory. In the EPR experiment a choice of measurements may be performed on the first system, and corresponding to this choice different observables in the second system will be assigned probability 1. Since Locality is assumed, no measurement performed on system 1 can disturb the separated system 2, and so Einstein argues from his second definition that each of the observables on system 2 represents an element of reality. However, since these are incompatible observables no quantum state can assign probability 1 to them all, and so Einstein concludes from his first definition that quantum theory is not complete.

1.2.4 Complementarity

The EPR paradox was specially designed by Einstein to show that a disturbance interpretation of quantum mechanics is not possible. However, Bohr responded to this paradox with a more sophisticated version of the disturbance interpretation which he subsequently developed into his theory of 'complementarity'. Bohr agreed that the Principle of Locality was desirable, especially as it seems to be required by relativity theory. But he argued that Einstein's definition of an element of reality, and hence also his definition of a complete theory, is at fault.

Bohr is notoriously difficult to understand, although he has written at length explaining his views[†]. Perhaps, however, these can be summarised briefly as follows. He claims that we should deduce from quantum theory that there exist 'complementary' elements of reality which by their very nature cannot be simultaneously assigned probability 1. The lack of classical quantum states stems from the new complementary things which the theory describes, rather than from an incompleteness in the theory. According to Bohr, the theory is not at fault, but instead we need (Bohr 1935 p 702)

a radical revision of our attitude, as regards physical reality.

Bohr defined complementarity in terms of measuring instruments. Two observables represent complementary phenomena if they must be measured by procedures which are physically incompatible. He analysed, for example, many experimental arrangements to show that any measurement of the position of an electron destroys essential knowledge of its momentum, and vice versa, so these phenomena are complementary. According to Bohr, the very meaning of the terms used in

---

[†] See for example Bohr (1958). Perhaps his clearest paper is *Discussion with Einstein on Epistemological Problems in Atomic Physics* in Schilpp (1969). Various papers have been written about Bohr. For example, there is a detailed examination of his arguments in Hooker (1972). An historical survey of Bohr's ideas, and of his debates with Einstein can be found in Jammer (1974), where many references to other works can be found.

quantum theory depends on the processes used to measure them. Thus (Bohr 1935 p 699):

> We have in each experimental arrangement suited to the study of proper quantum phenomena not merely to do with an ignorance of the value of certain physical quantities, but with the impossibility of defining these quantities in an unambiguous way.

Measurement disturbance is therefore important not simply because of its physical interference with the measured system, but also because it is 'built in' to the very definitions of our words. It follows that our descriptions of microscopic phenomena must always be limited. We must necessarily use macroscopic interfering measurement processes, and so our theory must necessarily use incompatible observables referring to complementary phenomena.

In fact, Bohr goes even further, and claims that this is the only possible way to generalise classical mechanics (Bohr 1935 p 700):

> In fact it is only the mutual exclusion of any two experimental procedures, permitting the unambiguous definition of complementary physical quantities which provides room for new physical laws, the co-existence of which might at first appear irreconcilable with the basic principles of science. It is just this entirely new situation as regards the description of physical phenomena that the notion of *complementarity* aims at characterising.

In their paper Einstein, Podolsky and Rosen anticipate one 'solution' to their paradox, which they then reject (Einstein *et al* 1935 p 780):

> One would not arrive at our conclusion if one insisted that two or more physical quantities can be regarded as simultaneous elements of reality *only when they can be simultaneously measured or predicted.* On this point of view, since either one or the other, but not both simultaneously, of the quantities P and Q can be predicted, they are not simultaneously real. This makes the reality of P and Q depend on the process of measurement carried out on the first system, which does not disturb the second system in any way. No reasonable definition of reality could be expected to permit this.

It seems in fact that it is just this 'unreasonable' solution which Bohr adopts. Phenomena are complementary according to Bohr, if they cannot be simultaneously measured, and in this sense he might admit that they may not be simultaneously 'real'. However, Bohr denies that his view leads to paradox. He does seem to agree that the choice of measurement performed on one system will determine which properties are real, i.e. which can be predicted with certainty, on the other, but he insists that this is not paradoxical. Even though he has agreed to the Locality principle, he insists that there is a sense in which even the separated system is disturbed by measurement (Bohr 1935 p 700) :

> Of course there is in a case like that just considered the EPR experiment no question of a mechanical disturbance of the system under investigation during the last critical stage of the measuring procedure. But even at this stage there is essentially the question of *an influence on the very conditions which define the possible types of predictions regarding the future behaviour of the system.*

We seem very far now from the attractively simple 'disturbance' interpretation discussed in the beginning of this section!

1.2.5 *The two-slit experiment*

The 'two-slit paradox' was often used by Bohr to illustrate his views on complementarity. In this experiment, a source of electrons is placed on one side of a diaphragm, on the other side of which is a photographic plate. There are two small slits in the diaphragm through which electrons can pass. If one slit is covered, and a beam of electrons is aimed at the diaphragm, a concentration of dark spots will appear on the plate directly opposite the hole, presumably each spot marking a 'hit' of an electron. When both slits are open in front of the electrons a different kind of pattern appears. In a corresponding macroscopic experiment one would find a simple 'sum' of the hits opposite each hole, as one would if each hole were uncovered in a separate experiment. However, one finds instead (in the microscopic two-hole experiment) a pattern on the photographic plate which shows interference effects characteristic of wave phenomena. This is quite unlike the analogous macroscopic experiment with particles fired through two holes.

The paradox in this experiment becomes clearer if the beam of electrons is reduced until the electrons are emitted one at a time, and the apparatus is left for a while to accumulate a pattern on the photographic plate. In this case we find, as quantum mechanics predicts, that it is the interference pattern which arises on the plate. Yet if one supposes that electrons are like macroscopic balls, then they must surely pass from the source through only one slit to the plate, and so a paradox arises, for one would expect the 'sum' of the hits concentrated opposite each hole. It seems that even when the electrons pass one at a time, the two-slit experiment differs from the two single-slit experiments. The existence of two open slits alters the behaviour of each individual electron.

In the last chapter of this book we shall discuss various ways in which this experiment is regarded as paradoxical. Here we shall simply note that Bohr often used this example to illustrate his views on complementarity (Schilpp 1969). He argued that the paradox arises from our assumption that each electron has a well-defined 'course' or trajectory from its source to the photographic plate. Yet such a course can in fact only be determined if both the position and the momentum of an electron is known as it passes through the diaphragm. Bohr analysed in great detail the different ways in which one might alter the apparatus of the experiment in order to measure these observables, and he concludes that simultaneous measurement is in fact not possible. For example, if the position of the electron at either slit is to be known exactly, then the diaphragm must be rigidly fixed and one slit covered, but when the diaphragm is fixed the momentum transferred from the electron to the diaphragm cannot be measured, and anyway when one slit is covered we know that the interference pattern disappears. On the other hand, if we measure the momentum transfer, for example by suspending the diaphragm on little springs attached to a scale, then the position of the diaphragm and its slits becomes uncertain. In such a way Bohr argues that simultaneous measurement of position and momentum is impossible, these observables are therefore incompatible, and hence it is a mistake to regard an electron as having both a position and momentum simultaneously. It follows that we cannot regard the electrons as having a well-defined trajectory in this experiment, and so no paradox about this trajectory can arise (Schilpp 1969 p 218):

> ...it is only the circumstance that we are presented with the choice of *either* tracing the path of a particle, *or* observing interference effects, which allows us to escape from the paradoxical necessity of concluding that the behaviour of an electron or a photon should depend on the presence of a slit in the diaphragm through which it could be proved not to pass. We have here to do with a typical example of how the complementary phenomena appear under mutually exclusive experimental arrangements ... and are just faced with the impossibility, in the analysis of quantum effects, of drawing any sharp separation between an independent behaviour of atomic objects and their interaction with measuring instruments which serve to define the conditions under which the phenomena occur.

Thus according to Bohr it makes no sense to ask about the 'course' of an electron when the interference effects appear, for this term can only be defined in terms of measuring processes which are incompatible with each other and with the observed effects. The paradox is defined out of existence, and cannot even be phrased.

1.2.6 Quantum logic

Although it is now more than fifty years since Bohr and Einstein first argued with each other about the interpretation of quantum theory, the essential issues remain the same. An enormous amount of work has since been done on the mathematical details of the theory, and much has been written about its interpretation, but little has been contributed which really helps our understanding. Since these arguments of the 1920s and 1930s there has probably been only one truly influential new interpretation of quantum mechanics. This is the so-called 'quantum logical' view, according to which quantum mechanics uses a logic all of its own. According to the 'quantum logicians', subatomic systems must be described using laws of logic which are drastically different from those which govern everyday classical descriptions. Although this view only became prominent in the 1960s, with the work of Jauch, Piron and others, it did in fact originate in the 1930s with the work of von Neumann and his collaborator Birkhoff.

In their paper *The Logic of Quantum Theory* Birkhoff and von Neumann (1936) consider an example which is often cited as a logical paradox of quantum theory. This will be discussed in detail in §7.3, but briefly the authors describe it as follows. We let symbols a, a' and b represent three different observations: a is to denote 'the experimental observation of wave-packet ψ on one side of a plane in ordinary space', a' corresponds to 'the observation of ψ on the other side', while b represents 'the observation of ψ in a state symmetric about the plane'. These authors then argue that the probability of finding both b and either a or a' is the same as finding just b, since a and a' exhaust the possibilities. However, the probability of finding both b and a is zero, as is the probability of finding both b and a'. Thus, they claim, a fundamental law of classical logic fails:

$$\text{b and (either a or a')} \neq \text{(b and a) or (b and a')}.$$

The probability assigned to the left-hand side is, they say, just the probability of b, which is assumed to be non-zero, yet the probability assigned to the right-hand side is zero. It follows, according to the authors, that these two expressions cannot be equivalent in the logic of quantum descriptions. Yet this expression gives an instance of the Law of Distribution, which is fundamental to classical logic.

Modern advocates of quantum logic argue along the lines suggested by von Neumann (1955) in the section of his book entitled *Projections as Propositions*, and this argument too will be examined in detail in §7.3. Here he discusses quantum properties such as those of the form 'Observable A has value a', and he argues that these, and more properties formed from these, are properly represented by the projection operators of the Hilbert Space representing the quantum theory. Von Neumann (1955) remarks that the relationship between projections and properties:

makes possible a sort of logical calculus with these.

Since there is a one-to-one correspondence between the subspaces of a Hilbert Space and its projection operators, it has often been claimed that the lattice of subspaces represents a 'quantum logic'. A great deal of mathematics has been done in the last twenty years, analysing the lattice properties of this structure and others like it. It is clear, at least, that such lattices can only represent a drastically non-classical logic.

The details of this argument will be examined later. However, it is hard to see how such a view could help us to understand quantum mechanics. If we think, argue, and understand using classical rules of logic, then what does it mean to say we must use quite different rules when discussing microscopic phenomena? How could this solve our conceptual problems? With all the mathematical activity, little attention has been spared for such issues, and yet how amazing it would be, and how important for any logician, if the quantum logicians were right!

1.3 Why use logic?

We have now seen some of the puzzling features of quantum mechanics, and indeed there are many more 'paradoxes' than those discussed here. However, the most fundamental problem is to understand quantum probability assignments, in particular to understand what the probabilities can tell us about individual systems. We would then understand why quantum states are irreducibly statistical, and why probability assignments 'jump' after measurements are performed.

In this book, the methods of modern logic will be applied to this problem. Logic, at least as it has developed this century, studies the ways in which propositions can consistently be true or false. Logic makes explicit the rules which are used to derive complex conclusions from the truth or falsity of simple propositions. Thus we shall understand the logic of quantum theory when we know firstly which simple propositions it uses to describe microscopic realities, and secondly when we know how these can be combined to form complex descriptions. The logic tells us, in this sense, how physical systems are described by a theory.

If, in addition to knowing the logic used by quantum theory, we also understand the relationship between probability assignments and this logic, then we will know which individual descriptions can be true or false when a probability assignment is made. Clearly this would solve the fundamental problem mentioned above. When the relationship between the logic and probabilities of a quantum theory is properly understood, then we will know what these probabilities have to say about individual systems.

The importance of logic to the problem of understanding quantum mechanics has been recognised before, and various attempts have been made to use logic to solve this problem. (For a concise review of these attempts, see for example Jammer 1974 ch 8.) Reichenbach for example, was noted for his early attempt to define compatibility as a logical operation in a three-valued propositional system (Reichenbach 1944), and as was mentioned in the last section, there has recently been a great vogue for 'quantum logic', supposedly a non-classical, non-distributive propositional logic which is represented by the lattice subspaces of Hilbert Space.

In this book a rather different kind of programme will be carried out. The fundamental issue here is the relation of logic to probability theory. It is often assumed that we fully understand the nature of this relation in classical mechanics, but I shall argue that this is not the case. A thorough reappraisal of the logical foundations of probabilities in mechanics will be carried out. After an introduction to the methods of modern logic (in chapter 2), I shall begin with a detailed account of the propositional logic used in any theory of mechanics, including classical and quantum theories (in chapter 3). This will lead to discussion of the logical characteristics of measurement, and of states (in chapter 4). After outlining the accepted principles of probability theory (in chapter 5), this logical analysis can be used to develop a new approach to probabilities in mechanics (in chapter 6). This will provide a clear understanding of quantum probabilities.

The discussion of logic in the next sections is self-contained, although brief. Where any section in the following chapters is particularly daunting, a 'Summing up' section is given at the end, which very briefly presents the key issues in a less technical way.

2 Modern Logic

2.1 Introduction to logic

The term 'logic' is used nowadays to mean very different things, and this ambiguity is especially evident in discussions of quantum theory. Mathematicians and physicists frequently use the term to mean no more than an algebra, or perhaps any partially ordered set or lattice, which has some connection with a theory. Thus, for example, the Hilbert Space representing a quantum theory is often called the 'logic' of that theory with no attempt at justification or discussion at all.

Here, however, we return to the traditional meaning of the term. Logic is traditionally the study of reasoning, of argumentation. Modern formal logic makes this study systematic. One considers firstly the ways in which simple propositions can be combined to form complex assertions, and secondly how the truth or falsity of simple propositions affects the truth-value of other complex statements in the system. In this way, one can represent and compare different systems of reasoning. Algebraic methods are certainly used in modern logic, as will be discussed in the next section, but an algebra is of interest only if it represents a system of propositions.

In this section the traditional methods of modern logic will be briefly discussed. Classical bivalent logic (CBL for short) will serve as an example. This is a simple propositional system which has important algebraic properties, as we will see later. A detailed treatment of this logic can be found in any modern elementary text on logic†.

2.1.1 Propositions

Any logic is generated from a fundamental set of simple propositions or propositional variables. Propositions are essentially statements which have the property of being sometimes true or false. Thus for example This house is green is a proposition, while 'Shut the door!' is not. (Philosophers sometimes wish to distinguish propositions from

† The logic we call CBL is variously called the Propositional Calculus (PC), the Sentential Calculus or simply Classical Logic. Examples of texts discussing this system are Hughes and Londey (1965) or for a more formal treatment see e.g. Mendelsohn (1964). There is a brief discussion of this system also in Hughes and Cresswell (1968 ch 1).

sentences—propositions are what sentences express. This distinction is not important here.) In a formal system of logic, such simple propositions will be represented by propositional variables p, q, r, ... An abstract logical system can simply be generated from the variables themselves.

Complex propositions will be constructed from simple ones by means of logical connectives. These vary according to the logic considered. Those used in classical logic are negation, represented by the symbol $\neg$, disjunction represented by $\lor$, conjunction $\cdot$, and hook $\supset$, which is used to express implications, i.e. complex propositions of the form If..., then .... Thus if This house is green is a proposition in a classical logic, represented by variable p, This house is not green, the negation of p, is also a proposition in the logic and will be represented by the formula $\neg$p.

Similarly, let us suppose that the propositions You are sitting and You are asleep are those of a classical logic, represented by variables p and q respectively. The complex formulae p $\lor$ q and p·q represent the disjunction You are sitting or you are asleep and the conjunction You are sitting and you are asleep respectively. The formula p $\supset$ q represents the complex proposition If you are sitting, then you are asleep. Normally only one or two connectives are taken as fundamental to a logic, and others are defined in terms of these. In classical logic it is usual to choose negation and hook, or else perhaps negation and disjunction.

Any logic must specify exactly how complex propositions can be formed from simple ones. For example, we must know that negation is a unary operator in CBL, and so symbol $\neg$ can only appear in front of a variable, and not for example between two variables. Thus $\neg$p is a well-formed string of symbols in the logic, while p $\neg$ q is not. Brackets are used to indicate on what part of an expression a connective is to act. For example, the formulae $\neg$(p $\lor$ q) and $\neg$p $\lor$ q represent different complex propositions. In the first case, the brackets indicate that the connective is applied to the entire expression p $\lor$ q, while in the second this connective is applied only to p. For example, suppose that p represents It is raining and q It is snowing. Then the formula $\neg$(p $\lor$ q) represents It is neither raining, nor is it snowing, while $\neg$p $\lor$ q stands for It is either not raining, or it is snowing, which is clearly quite a different proposition. Similarly, the following are all unambiguous formulae in a classical logic: p $\supset$ (q $\lor$ r), $\neg$(p $\supset$ r), and p $\supset$ (q $\lor$ (r·p)).

Hence any formal system of logic must specify firstly an alphabet of symbols, which in the case of classical logic include the propositional variables and logical connective symbols. Secondly it must specify rules for combining the symbols into meaningful strings, or 'well-formed formulae', known as 'wffs' for short. Thus in the formal definition of classical propositional logic we would include:

Alphabet of CBL

p, q, r, . . .	propositional variables
⌐, ⊃	logical connectives
(,)	brackets

Rules of Formation for CBL

1 Any propositional variable is a wff.
2 If α is a wff, so is $\neg\alpha$.
3 If α, β are wffs, so is $\alpha \supset \beta$.
4 Only strings generated by rules 1–3 above are wffs.
   Normal conventions of bracketing apply.

Here the connectives $\neg$ and $\supset$ are taken as fundamental. Once the meaning of these connectives has been defined, the other connectives can be introduced as abbreviations (in the next section). From now on the letters $\alpha, \beta, \gamma, \ldots$ are used to refer to wffs in a logic, while p, q, r, . . . are retained specifically as simple propositional variables.

2.1.2 Valuations

So far we have discussed the symbols which make up the alphabet of a logic, and the rules showing how these symbols can be combined to form well-formed formulae. Nothing so far has determined the meaning of the connectives, and hence the meaning of complex formulae. In fact, to say that these are *logical* connectives is just to say that they are defined solely by 'truth conditions'. The truth or falsity of a complex proposition depends solely on the truth or falsity of constituent simple propositions. The rule which determines this dependence is the rule which defines the logical connective.

Intuitively it is easy to understand the meaning of logical connectives. For example, negation in classical bivalent logic is defined by the condition that wff $\neg\alpha$ is true only when the constituent wff α is false, and $\neg\alpha$ is false only when α is true. This definition does express a common meaning of the term 'not' in everday English. This house is not green will only be true when it is false that This house is green and vice versa. Similarly connective · is defined in classical logic by the condition that a conjunction $\alpha \cdot \beta$ is true only when both the conjuncts α, β are true, and false if either is false. This definition also expresses common usage, in this case of the word 'and' in everyday English. We say that You are sitting and you are asleep is a true complex proposition only if it is both true that You are sitting, and true that You are asleep. If in fact you are standing or wide awake, so that one of the simple constituents is false, then the whole complex proposition is also false.

To express these conditions in a formal and accurate way, one must introduce the notion of a valuation. This is an assignment of truth-values to propositions. It is a function from the set of simple propositions or propositional variables of a logic to another set which contains at least the two truth-values 'true' and 'false'. If we represent these by the symbols t and f respectively, then we can define for classical bivalent logic:

Definition
A *(simple) valuation of* CBL is a function h, $h: \{p, q, r, \ldots\} \rightarrow \{t, f\}$ which maps every propositional variable onto one of the two truth-values.

A logic is *bivalent* when the range set of its valuations contains only two values. That is, every valuation assigns each proposition to one of the two truth-values. In other cases, we say a logic is *many-valued*, or *n-valued* where n is the number of elements in the range of a valuation.

The conditions defining logical connectives extend any simple valuation of a logic to a valuation of complex propositions. For example, consider negation in classical logic. According to the condition discussed above, the negation of any wff, $\neg\alpha$, is to be true whenever α is false, and vice versa. This can be expressed as the following valuation rule:

Valuation Rule for $\neg$ in CBL
1   If $h(\alpha) = t$, then $h(\neg\alpha) = f$
2   If $h(\alpha) = f$, then $h(\neg\alpha) = t$

A simple valuation h assigns truth-values t and f to simple propositions. The rule defining $\neg$ in CBL extends h to a function which also assigns wffs of the form $\neg p$ to the two truth-values.

Clearly, a valuation will include all wffs in its domain if we introduce a rule defining hook as well as one defining negation, since these two connectives were taken as the fundamental. Hook is slightly less intuitive than either negation or conjunction, simply because it does not coincide so well with a word or phrase in the English language. As remarked earlier, it is used to express implications, and can be represented by the phrase If . . . , then . . . . In fact it is defined so that an implication fails, i.e. a wff of the form $\alpha \supset \beta$ is false, only when the antecedent α is true and the consequent β false. In every other case wff $\alpha \supset \beta$ will be true. This does to some extent reflect ordinary usage. For example, consider the complex proposition If someone is there, then it's Mabel, represented in CBL by the wff $p \supset q$. According to the definition, this wff will be false only if it is true that Someone is there, but false that It is Mabel, and this seems to coincide with everyday language. However, in the case where no-one is there, and the antecedent p is false, then the implication $p \supset q$ is true according to the definition CBL employs. This may seem peculiar, but it should be remembered that CBL is a relatively unsophisticated logic, and the hook is designed primarily to exclude implications which lead from truth to falsity. So long as no false proposition can be derived from a true one in the logic, one can be satisfied that a useful sense of implication is expressed. This logic does not attempt to express all the subtleties of the English language.

The conditions defining hook are expressed in the following valuation rule:

Valuation Rule for $\supset$ in CBL
1   If $h(\alpha) = $ t and $h(\beta) = $ f, then $h(\alpha \supset \beta) = $ f
2   $h(\alpha \supset \beta) = $ t otherwise

By condition 2, when either $h(\alpha) = $ f, or else both $h(\alpha) = $ t and $h(\beta) = $ t, then $h(\alpha \supset \beta) = $ t.

Since only negation and hook were taken as fundamental in forming the wffs of classical logic, these two valuation rules will define all its connectives. Hence the rules extend any simple valuation h assigning truth-values to simple propositions, to a unique complex valuation that assigns truth-values to any wff in the system.

The valuation rules can be expressed very easily using truth tables. These set out all the possible valuations of constituent propositions in a wff, giving the corresponding valuation of complex wffs under the connective symbol. For example, we can represent the two valuation rules given above on the following tables:

$\neg$	α
f	t
t	f

α	$\supset$	β
t	t	t
t	f	f
f	t	t
f	t	f

In the first table, all possible values of the constituent wff α are included in a column underneath it (recall that in logic CBL any wff must be either true or false in any valuation). The corresponding value of the complex wff $\neg\alpha$ is recorded under $\neg$. Thus the table tells us that when α is true, $\neg\alpha$ is false, and when α is false, $\neg\alpha$ is true, as the valuation rule for $\neg$ specified. In the table defining $\supset$, we need four rows to record all possible combinations of valuations of the two constituents. (In general 2^n rows will be required to determine all valuations of a wff with n constituent variables.) Row 1, for example, tells us that when α is true and β is true, wff $\alpha \supset \beta$ is true, while by row 2 we know that when α is true but β is false then the implication is false, and so on.

Other connectives in the logic can be introduced as abbreviations for strings of primitive symbols. It is usual in CBL to introduce disjunction $\vee$, conjunction $\cdot$ and the biconditional $\equiv$. The following definitions can be used for these connectives:

Abbreviations for CBL

$$\alpha \vee \beta = (\alpha \supset \beta) \supset \alpha$$
$$\alpha \cdot \beta = \neg(\neg\alpha \vee \neg\beta)$$
$$\alpha \equiv \beta = (\alpha \supset \beta) \cdot (\beta \supset \alpha)$$

The properties of these connectives can be summarised in the following truth table:

α	β	$\alpha \cdot \beta$	$\alpha \vee \beta$	$\alpha \equiv \beta$
t	t	t	t	t
t	f	f	t	f
f	t	f	t	f
f	f	f	f	t

These values can be checked by constructing truth tables for the expressions on the right-hand side of the abbreviationary definitions. The tables show that, for example, conjunction has the properties already discussed. A conjunction is true only when both conjuncts are true. Similarly disjunction turns out to have the property that $\alpha \vee \beta$ is true when either α or β is true, and false only when both are false. This coincides fairly well with normal usage of the English word 'or'. If p represents He bought a paper and q He bought a book, then the complex proposition He bought a paper or a book represented by p ∨ q, is true when either p or q is true. The only slight difference from ordinary usage is that according to CBL the complex proposition is also true when both disjuncts are true, so when he bought both a paper and a book, wff p ∨ q is true. Lastly, connective ≡, the biconditional, is used to express logical equivalence. According to the definition, wff $\alpha \equiv \beta$ is true only when the two constituents have exactly the same truth-value. This connective is particularly useful in discussions of algebra, as we shall see later.

2.1.3 The structure of a logic

We have now seen how a propositional logic is defined. An alphabet, which includes propositional variables and connective symbols, is specified, as are rules of formation for combining these symbols. Logical connectives are defined by valuation rules, which uniquely associate any simple valuation with a complex valuation assigning truth-values to all wffs of the system. Lastly, abbreviations are specified for any extra connectives. To illustrate, we have given the alphabet, rules of formation, valuation rules and abbreviations defining classical bivalent propositional logic.

Different logics will vary from CBL in at least one of the first three respects. That is, a different logic will have a different alphabet, a different set of wffs, or will use different conditions to define the connectives. Any logic which includes in its alphabet only propositional variables and logical connectives, apart from brackets, is called a *propositional* logic. Other kinds of logic may include other symbols. For example in a *predicate* logic one analyses propositions into constituent parts, and so such a logic may include predicate variables, noun variables and quantifiers in its alphabet (see, for example, Hughes and Londey 1965). Another logic may be propositional and have the same wffs as CBL but define the connectives differently. In principle, one could have as many such logics as there are different conditions for defining $\neg$ and $\supset$, although such systems may actually have little interest as systems of reasoning. Later we shall discuss *modal* propositional logics, which include all the symbols and wffs of an ordinary propositional system, but which in addition include an operator which is not defined in a truth-functional way—the rule defining a modal operator cannot be expressed in a truth table.

The tautologies or valid formulae of a logic are the wffs which are true in every valuation.

Definition
α is a *tautology of CBL* (or *valid*, or a *logical truth*) if $h(\alpha) = t$ for every valuation h of CBL.

These wffs characterise the structure of a logic in the sense that they are true by virtue of their form and the definitions of connectives, rather than by virtue of the particular truth-values of their constituent propositions. In this sense they express the properties of a logic.

Consider for example some tautologies of CBL. The wff $\neg\,\neg\alpha \equiv \alpha$ is a logical truth, traditionally called the 'Law of Double Negation'. We can check this by constructing a truth table:

$\neg$	$\neg$	α	$\equiv$	α
t	f	t	t	t
f	t	f	t	f

In any valuation the wff is true, so it expresses a property of negation in this logic. The truth of the wff stems solely from the conditions defining connectives. Similarly, wff $\alpha \lor \neg\alpha$ is a tautology of CBL, known as the 'Law of Excluded Middle'. Its validity follows from the definition of $\lor$, and of a valuation in CBL as a function taking every proposition to one of the two truth-values. Again one can check using a truth table:

α	$\vee$	$\neg$	α
t	t	f	t
f	t	t	f

In every valuation the wff is true, and so it is a tautology of this bivalent logic CBL.

Symbols can be used to express the relations between any wff and any valuation of a logic. Since CBL is bivalent, there are just two such relations for this logic. Either a wff is true in a valuation h, or it is false in that valuation. The 'double turnstile' symbols $\models$ and $\dashv$ can be used to express this relation, with a subscript to indicate the valuation concerned.

Definition
CBL $\models_h \alpha$ iff $h(\alpha) = $ t (h a valuation of CBL)
CBL $\dashv_h \alpha$ iff $h(\alpha) = $ f

In this case, $\models$, $\dashv$ are three-place relations between logic, valuation and wff. However, it is usual to omit reference to the logic where this is understood, so that usually one writes, for example, $\models_h \alpha$. Similar 'double turnstiles' without subscripts are used to indicate the logical truths and falsehoods of a logic. For example:

Definition
CBL $\models \alpha$ iff CBL $\models_h \alpha$ for every valuation h of CBL
CBL $\dashv \alpha$ iff CBL $\dashv_h \alpha$ for every valuation h of CBL

In the first case we say α is a *tautology*, or is *valid*, or is a logical truth, while in the second α is a *logical falsehood*. A stroke through any symbol indicates that the relation fails to hold. For example, $\not\models_h \alpha$ indicates that α is not true in h, and so $h(\alpha) \neq$ t. Similarly $\not\models \alpha$ means that α is not valid, and so $h(\alpha) \neq$ t for some valuation h. Again one can omit reference to the logic where this is understood.

We have seen that the double turnstiles express relations between wffs and valuations. With subscripts, they tell us the value of a wff in that valuation and without subscripts they tell us if a wff is valid or is logically false in a logic, thus relating the wff to the whole set of valuations of a system. These symbols should not be confused with similar 'single turnstiles', which will be briefly discussed in the next section. We shall see below that the single turnstile does not express a relation between wffs and valuations, but instead concerns the derivability of a wff in some axiomatic system.

2.1.4 Axiomatics

The definition of a logic discussed here is not the only one. Instead, one can regard a logic as an axiomatic system, in which case it must be defined rather differently. Instead of specifying an Alphabet, Rules of Formation, and Valuation Rules, a Logic is defined by specifying an Alphabet, Rules of Formation and a set of special wffs called Axioms, as well as Rules of Deduction. These deduction rules allow one to generate more wffs from the axioms. All such derived formulae are called theorems. The axiomatic approach identifies a logic with its set of theorems.

Although the axiomatic method will not be used later, it is perhaps interesting to represent the logic CBL as an axiomatic system. The first part of such a definition coincides with that given earlier (in §2.2.1).

Alphabet of CBL

p, q, r, . . .   propositional variables
$\neg$, $\supset$ logical connectives
(,)   brackets

Rules of Formation for CBL

1   Any propositional variable is a wff.
2   If α is a wff, so is $\neg\alpha$.
3   If α, β are wffs, so is $\alpha \supset \beta$.
4   Only strings generated by rules 1–3 above are wffs.
    Normal conventions of bracketing apply.

Instead of the Valuation Rules, which were earlier used, we now specify a set of Axioms and the Rules of Deduction.

Axioms for CBL

A1 $(\alpha \supset (\beta \supset \alpha))$
A2 $((\alpha \supset (\beta \supset \gamma)) \supset ((\alpha \supset \beta) \supset (\alpha \supset \gamma)))$
A3 $((\neg\beta \supset \neg\alpha) \supset ((\neg\beta \supset \alpha) \supset \beta))$

Rules of Deduction for CBL

MP   (Modus Ponens):   If α and $(\alpha \supset \beta)$ are theorems, so is β.
US   (Uniform Substitution):   The result of uniformly replacing any variable in a theorem is itself a theorem.

We have now defined the axiomatic system CBL: it is the set of theorems derivable from the axioms by means of the two Rules of Deduction. Strictly speaking, a theorem is established only when it is shown to be the last line of a proof, or derivation, each line of which must be either an axiom, or be derivable from either an axiom or an earlier line, by one of the two Rules. A full proof is generally very cumbersome and in practice one introduces abbreviations, derived rules of deduction and other short-cuts to discover theoremhood. However, the following simple example of a proof may be of interest. This establishes the classical

'Law of Identity' as a theorem of CBL. The steps are numbered on the left, with their justification noted on the right.

1 $\vdash (\alpha \supset ((\alpha \supset \alpha) \supset \alpha))$
   $\supset ((\alpha \supset (\alpha \supset \alpha)) \supset (\alpha \supset \alpha))$ From Axiom A2 by US
2 $\vdash \alpha \supset ((\alpha \supset \alpha) \supset \alpha)$ From Axiom A1 by US
3 $\vdash (\alpha \supset (\alpha \supset \alpha)) \supset (\alpha \supset \alpha)$ From 1, 2 above by MP
4 $\vdash \alpha \supset (\alpha \supset \alpha)$ Axiom A1
5 $\vdash \alpha \supset \alpha$ From 3, 4 above by MP

This shows that $\alpha \supset \alpha$ is indeed a theorem of CBL.

Notice that the single turnstile symbol $\vdash$ has been used at the beginning of each line. This symbol indicates that the expression following is a theorem of the system CBL. In fact, the single turnstile is a relational symbol, which should strictly speaking stand between a logic and a wff, so for example, line 4 above should read CBL $\vdash \alpha \supset (\alpha \supset \alpha)$. (Notice that, in fact, every line of any proof will be a theorem, since it will have the preceding lines as a proof. Of course, since A1 is an axiom, it is actually an example of a one-line proof.) Thus the single turnstile is used to express the relation of a wff to an axiomatic system, while the double turnstile expresses the relation of a wff to a valuation or to a set of valuations of a logic.

There is a connection between the two different definitions of a logic. Much of the work of traditional logicians, especially early this century, consisted in demonstrating that some logic defined in the earlier manner, with rules of formation and valuation rules, could also be represented axiomatically. To do this, one must first show that the tautologies of the logic coincide with the theorems of some axiomatic system. That is to say, one shows that when $L \models \alpha$ for some wff α and logic L, then also $L' \vdash \alpha$ for an axiomatic logic L'. This property is called *completeness* of the logic L. Second, one shows the converse, that every theorem of the axiomatic system is also a tautology of the logic, i.e. that when $L' \vdash \alpha$, then $L \models \alpha$, a property called the *soundness* of the logic L. When both these results are established one has shown that the logic and axiomatic system are equivalent. Not every logic is sound or complete.

It can be shown that every theorem of CBL, derived from the axioms by the rules of deduction, is a tautology of the logic system defined earlier, and, vice versa, every tautology is a theorem of the axiomatic system. Hence CBL is both sound and complete.

2.1.5 Summing up

A formal system of propositional logic is defined in at least three stages. First an *Alphabet* of symbols is specified, which will contain propositional variables and logical connectives. The variables may represent simple propositions. Secondly *Rules of Formation* must be given, so that the symbols in the alphabet can be combined to form well-formed formulae (wffs) of the logic. These represent complex propositions. Thirdly *Valuation Rules* will define the logical connectives in a truth-functional way. That is, the rules extend any truth-value assignment to simple propositions to a unique assignment of truth-values to complex wffs. It is also usual to introduce abbreviations which define derived connectives. In the discussion above it was also noted that a logic can be defined in a different way, as an axiomatic system.

In this section, classical bivalent logic (CBL) was discussed, and used to illustrate the methods of modern logic. In chapter 3 these methods will be applied to mechanical theories, for we discuss there the propositional logic which is used by any mechanical theory to express its descriptions of reality. In the remaining sections of this chapter we introduce algebraic methods used in the study of logic, which will be useful in later discussion.

2.2 Algebraic methods

2.2.1 Introduction to lattices

Before explaining how modern algebra can be applied to the study of logic it is necessary to review the principles of lattice theory†.

Fundamental to algebra is the notion of a *set* of elements, $A = \{a, b, c, \ldots\}$ and of relations between members of the set. The set A and its relations make up a relational *structure* $A = \langle A, \{R_i\} \rangle$, where $\{R_i\}$ is a family of relations on the set A. Intuitively, sets and relations are easy to understand, as we are familiar with them in everyday life. For example, our own family is a set whose elements are people, and they have relations defined among them. One member will be the mother of several others, one may be the brother of another and so on. Formally, we represent relations by sets of ordered *n*-tuples. For example the relation 'is the mother of' would be represented by those ordered pairs of elements in the set for which this relation holds. If we know that the pair (Sheila, John) is a member of this relation, then we known that Sheila is the mother of John.

Formally, a *relation* is defined as a subset of some cross-product. The *binary cross-product* of set A, $A \times A$ or A^2 is the set of ordered pairs of elements of A, $A^2 = \{(a, b) : a, b \in A\}$. In general an *n-ary cross-product* A^n is the set of ordered *n*-tuples of elements of A, $A^n = \{(a_1, \ldots, a_n) : a_i \in A, 1 \leqslant i \leqslant n\}$. A *binary relation* R is a subset of A^2 (i.e. $R \subseteq A^2$) and in general an *n-ary relation* R_n is a subset of the *n*-ary cross-product A^n, $R_n \subseteq A^n$. Thus a binary relation is just a set of ordered pairs of elements of A, while an

† The fundamental text on lattice theory is Birkhoff (1967). There are now many other works on the subject, including Gratzer (1971).

n-ary relation is a set of ordered *n*-tuples. It is convenient in the case of a binary relation R if we write $a\,R\,b$, instead of the more awkward $(a, b) \in R$.

Some properties of relations are particularly interesting to an algebraist. For example, the following are important properties of some binary relations. If R is a binary relation on set A, then

Definition
R is *reflexive*   if $a\,R\,a$ for every *a* in A.
R is *antisymmetric* if, whenever $a\,R\,b$ and $b\,R\,a$, then $a = b$.
R is *transitive*   if, whenever $a\,R\,b$ and $b\,R\,c$, then $a\,R\,c$.

An example of a relation which has all three properties is the relation of 'less than or equal to', $\leqslant$, on any set of real numbers. For instance, this is reflexive since every real number is equal to itself, and similarly one can check that it is antisymmetric and transitive. As we shall see later, these properties are also interesting to the logician, since the first and third at least are usually assumed to be properties of logical inference.

These three properties are of such interest in algebra that a special name is given to any set on which such a relation is defined:

Definition
If relation R on A is reflexive, antisymmetric and transitive, then R is a *partial ordering* of A, and is represented by symbol $\leqslant$. In this case the structure $A = \langle A, \leqslant \rangle$ is called a *poset* (short for partially ordered set).

Note that the symbol $\leqslant$ is used quite generally to represent any partial ordering, and does not refer solely to the familiar 'less than or equal to' relation among real numbers.

Operations may be defined over a set of elements. Formally, an *operation* is a mapping from some cross-product of A, A^n, into A. For example, a unary operation $'$ associates with any member *a* of A some unique element a' of A, and a binary operation $\circ$ associates with any ordered pair of elements (a, b) a uniquely defined element $a \circ b$. For instance, we are familiar with arithmetic operations on the real numbers. Negation can be viewed as a unary operation taking every number to its negative, and addition is a binary operation associating with every pair of numbers (x, y) the uniquely defined number $x + y$. In fact, one can regard any operation as a special kind of relation. Any unary operation $'$ corresponds to a binary relation consisting of all the pairs (a, a'), any binary operation $\circ$ corresponds to a 3-ary relation consisting of the triples $(a, b, a \circ b)$, and in general any *n*-ary operation corresponds to an $(n + 1)$-ary relation. The final element of the $(n + 1)$-tuple is uniquely determined by the previous *n* elements.

An *algebra* is a structure whose relations have the special property of being operations†. Thus an algebra A can be defined as a structure $\langle A, \{O_i\} \rangle$ where $\{O_i\}$ is a family of operations on A. Set A is called the *carrier set* of the algebra. Since the operations are defined over A, i.e. they map A^n into A, an algebra is closed with respect to its operations. Any subset B of A which is closed with respect to the operations of A is called a *subalgebra* of A. If the operations O_i are only defined for some elements of A and not for all elements, we may call A a *partial algebra*.

It is possible to define special elements in a poset and hence to introduce operations into such a set. First one must introduce the notion of boundedness. If B is some subset of A, then B is called *bounded* when the following elements exist:

Definition
a is an *upper bound* of B, if $a \in A$, and $b \leqslant a$ for every b in B.
a is a *lower bound* of B, if $a \in A$, and $a \leqslant b$ for every b in B.

An upper bound lies 'above' every element of the set B, while a lower bound lies 'below' each element, with respect to the partial ordering. One can now define the least upper bound of B, lub B, and the greatest lower bound, glb B, as follows:

Definition
a is the *lub* B if a is an upper bound, and for any other upper bound a', $a \leqslant a'$. Similarly, a is the *glb* B if a is a lower bound, and for any other lower bound a', $a' \leqslant a$.

These are the 'smallest' upper bound and 'largest' lower bound of the set respectively.

It is now possible to define two important operations called join and meet. These operations associate with two elements, their least upper and greatest lower bounds respectively:

Definition
The *join* of a, b, $\quad a \vee b = \text{lub } \{a, b\}$
The *meet* of a, b, $\quad a \wedge b = \text{glb } \{a, b\}$

One can check that the operations are well-defined by showing that these elements are unique, where they exist. In fact, this follows from the antisymmetry of the relation $\leqslant$. Clearly, in general, in an arbitrary poset A, the operations may not be defined everywhere, for subsets of A may not be bounded, let alone have a lub and glb. However, those posets which have this property are of especial interest and hence are given a special name:

---

† Sometimes this is called an *algebraic structure*, as it may also be required that an algebra includes two binary operations. The definition given here is probably the most general.

Definition

A *lattice* is a poset in which join and meet are defined everywhere.

Thus in a lattice, $a \vee b$ and $a \wedge b$ exist for any pair of elements a, b. In fact, an equivalent condition is that the lub and glb must exist for every finite subset B (see Gratzer 1971 p 4).

The operations $\vee$ and $\wedge$ have some interesting properties. These the following:

Lemma 2.1

In any poset A, where join and meet exist, they have the following properties:

L1	(Idempotence)	$a \vee a = a,\ a \wedge a = a$
L2	(Commutation)	$a \vee b = b \vee a,\ a \wedge b = b \wedge a$
L3	(Associativity)	$a \vee (b \vee c) = (a \vee b) \vee c$
		$a \wedge (b \wedge c) = (a \wedge b) \wedge c$
L4	(Absorption)	$a \vee (a \wedge b) = a \wedge (a \vee b) = a$

Moreover, $a \leqslant b$ is equivalent to each of the conditions:

$$a \vee b = b \qquad \text{and} \qquad a \wedge b = a.$$

In fact these properties fully characterise a lattice, in the sense that the following result holds:

Theorem 2.1

If $A = \langle A; \vee, \wedge \rangle$ is an algebra in which $\vee$ and $\wedge$ satisfy the conditions L1–L4 of lemma 2.1 above, then A is a lattice.

Thus if poset $A = \langle A; \leqslant \rangle$ is a lattice, it is equivalent to the algebra $A = \langle A; \vee, \wedge \rangle$. The two structures are the same (Birkhoff 1967 p 8).

We shall now discuss some different kinds of lattice, which will be of interest in later chapters. It can be shown (Birkhoff 1967 p 9) that any lattice at all will satisfy the following distributive inequalities:

$$\begin{aligned} \text{L5} \quad & a \vee (b \wedge c) \leqslant (a \vee b) \wedge (a \vee c) \\ & (a \wedge b) \vee (a \wedge c) \leqslant a \wedge (b \vee c). \end{aligned}$$

However, it is only in some special lattices that the corresponding identities hold:

Definition

Lattice A is a *distributive lattice* if the following distributive identities D1 and D2 hold for any elements a, b, c in A:

$$\begin{aligned} \text{D1} \quad & a \vee (b \wedge c) = (a \vee b) \wedge (a \vee c) \\ \text{D2} \quad & a \wedge (b \vee c) = (a \wedge b) \vee (a \wedge c). \end{aligned}$$

Not every lattice is distributive, but this is a property shared by many important algebras.

In any subset of a lattice, one can associate with each element a relative complement. This is defined as follows. If B is a subset of lattice A, and 1_B and 0_B represent the lub and glb of B respectively, then

Definition
a' is a *relative complement* of a in B (or alternatively a' is the complement of a relative to B), if

$$a \vee a' = 1_B \qquad \text{and} \qquad a \wedge a' = 0_B.$$

If we let 1 and 0 represent the universal bounds of a lattice A, so that 1, 0 are the lub and glb of lattice A respectively, then one can define an 'absolute' complement of each element, i.e.

Definition
a' is a *lattice complement* of a if

$$a \vee a' = 1 \qquad \text{and} \qquad a \wedge a' = 0.$$

Sometimes a' is simply called a *complement* of a in A. A lattice is said to be *relatively complemented* if every element has a relative complement in every subset containing it. A lattice is *complemented* if every element has a lattice complement. It is worth noting that the universal bounds 1 and 0 above coincide with the unit and zero elements of the algebra A. That is, the following conditions hold (Birkhoff 1967 p 9):

Lemma 2.2

$$a \vee 1 = 1, \qquad a \wedge 1 = a$$
$$0 \vee a = a, \qquad 0 \wedge a = 0.$$

In any distributive lattice, complements have some special properties:

Lemma 2.3
In any distributive lattice A, if element a has a lattice complement a', then it is unique.

It follows that in a distributive lattice we can regard complementation as a unary operation, associating with each element its unique lattice complement (Birkhoff 1967 p 17). Some properties of this operation are now presented.

Lemma 2.4
In a distributive lattice A the operation ' of lattice complementation satisfies the following conditions for any a in A:

C1 (Complement) $a \vee a' = 1, a \wedge a' = 0$
C2 (de Morgan) $(a \vee b)' = a' \wedge b', (a \wedge b)' = a' \vee b'$
C3 (Involution) $(a')' = a$

Any operation satisfying conditions C1–C3 is called an *orthocomplement*.

Of particular interest are complemented distributive lattices, i.e. lattices which are distributive and in which every element has a lattice complement. These are known as *Boolean lattices*, or *Boolean algebras*, in honour of George Boole (1815–64) whose work inspired interest in these systems.

Definition
A *Boolean lattice* is a distributive and complemented lattice.

Alternatively this could be defined by the properties of the operations on the system:

Definition
A *Boolean algebra A* is an algebra $A = \langle A; \vee, \wedge, ' \rangle$ where operations $\vee$ and $\wedge$ satisfy conditions L1–L4 above, as well as D1 and D2, and where operation $'$ satisfies C1–C3 above.

The equivalence of these two definitions is established in the following result:

Theorem 2.2
Every Boolean lattice is also a Boolean algebra, and vice versa, every Boolean algebra is also a Boolean lattice.

Boolean algebras have proved to be very interesting structures with many diverse applications. For example, naive set theory generates a Boolean lattice, since set inclusion is a lattice partial ordering, and set complements can always be defined. Alternatively, operations $\cup$ and $\cap$ of set union and set intersection satisfy the conditions L1–L4, D1 and D2, while set complement, $-$, satisfies conditions C1–C3. Hence set theory provides an example of a Boolean algebra. As we shall see in the next section, Boolean operations also have applications in the representation of propositional logic.

2.2.2 *The algebraic representation of a logic*

We now see how algebraic structures can represent propositional logics. The representation is needed because a logic is a complicated kind of system with a vast number of different wffs in its vocabulary. In order to analyse a logic, and to compare it with others, it is useful to use a simpler structure. Thus we wish to represent the wffs of a logic by a smaller set of elements, and to let operations and relations on this set represent logical connectives.

The usual method for constructing an algebra to represent a logic is called the Lindenbaum–Tarski method, after the two men who first developed it for classical bivalent systems. The basic idea is that all wffs which are logically equivalent, i.e. which always have the same value in any valuation, are represented by a single element of an algebra. They form a single 'equivalence class'. Logical connectives are represented by operations and relations among these elements. In this way a very much simpler structure can be used to represent any propositional logic. We now give the full construction of an algebra to represent CBL, the classical bivalent logic discussed in §2.1†.

Before constructing the algebra, one must first show that logical equivalence in CBL is indeed an equivalence relation in the accepted mathematical sense, which is to say it must be reflexive, transitive and symmetric. The first two properties were defined in the last section, but we repeat them here. For binary relation R on a set A:

Definition
R is *reflexive* if a R a for all a in A
R is *symmetric* if whenever a R b, then b R a
R is *transitive* if whenever a R b and b R c, then a R c
R is an *equivalence relation* if R is reflexive, symmetric and transitive.

A familiar example of an equivalence relation is the symbol $=$ on the set of real numbers. For instance, this is symmetric, since whenever $a = b$, then $b = a$ also. One can easily check that it is also reflexive and transitive.

Given that a relation R on some set A is indeed an equivalence relation, one can use to it 'partition' the set into mutually exclusive 'equivalence classes', which are just subsets of A, all of whose members are equivalent.

Definition
The *equivalence class* of a with respect to relation R, is

$$[a]_R = \{b \in A : aRb\}.$$

This is the set of all elements in A which are related to a by R. (Where a particular relation R is irrelevant or is understood, the subscript will be omitted.) The following are some important properties of equivalence classes.

Lemma 2.5
1   $a \in [a]$
2   If $b \in [a]$, then $[b] = [a]$
3   If $[a] \cap [b] \neq \varnothing$, then $[a] = [b]$

By 1, any element is contained in its own equivalence class. By 2, when any element is contained in the equivalence class of another, the two elements share the same equivalence class. Lastly, if two equivalence classes have any element in common they have all members in common. It follows that two different classes will always be disjoint.

---

† Full discussions of the method for classical bivalent logic can be found, for example, in Bell and Slomsen (1969 ch 2) and in Sikorski (1969).

As noted earlier, the equivalence relation of particular interest here is logical equivalence. We wish to partition the set of wffs in a logic into equivalence classes of wffs which have the same value in every valuation. Consider for example classical bivalent logic, CBL. In this logic the connective $\equiv$ is characterised by the condition that wff $\alpha \equiv \beta$ will be true only if both α and β share the same truth-value. If this wff is valid, i.e. true in every valuation, then it follows that α and β always have the same truth-value. Thus the appropriate relation expressing logical equivalence of wffs in this logic is

$$\alpha \, R \, \beta = \models \alpha \equiv \beta.$$

To show that this does indeed partition the wffs into mutually exclusive classes, one must show:

Lemma 2.6
Relation R above is an equivalence relation.

This result follows from the properties of the logic CBL. For instance, one can show that R is reflexive, by constructing a truth table to show that wff $\alpha \equiv \alpha$ is valid. Similarly, R is symmetric because if $\models \alpha \equiv \beta$ in CBL then $\models \beta \equiv \alpha$ also. Transitivity follows from a similar argument.

Having shown that R above is an equivalence relation, one can now construct the equivalence classes. It follows from the definition of R that these will be of the form

$$[\alpha] = \{\beta \in \text{CBL} : \models \alpha \equiv \beta\}.$$

An equivalence class contains all logically equivalent wffs. We shall denote the set of all equivalence classes by symbol CBL $+$. Thus

$$\text{CBL} + = \{[\alpha] : \alpha \text{ is a wff of CBL}\}.$$

So far, we have seen how a set of equivalence classes can be constructed to represent the wffs of a logic. One must now introduce relations and operations onto this set to represent logical connectives. Since $\neg$ and $\supset$ were taken as fundamental to this logic, it is these connectives which will be represented.

Recall that connective $\supset$ in CBL is a binary connective which is used to express propositions of the form 'If . . ., then . . .', i.e. $\supset$ expresses the implications of the logic. In the representative algebra we wish to analyse $\supset$; in particular we wish to represent implications which are logically true. This suggests that we define the representative relation on CBL $+$ as follows:

$$[\alpha] \leqslant [\beta] \qquad \text{iff} \qquad \models \alpha \supset \beta$$

i.e. the relation $\leqslant$ holds among equivalence classes if and only if there is a valid implication between represented wffs. Similarly, we wish to analyse properties of negation, $\neg$, in CBL, and so define a representative operation $'$ on CBL+

$$[\alpha]' = [\neg\alpha]$$

i.e. the complement of any equivalence class is the class which contains the negation of its representative wff.

Before discussing properties of the representative relation $\leqslant$ and the operation $'$, one must first show that they are indeed well-defined. That is, one must show that their definition does not depend on the choice of wff representing a class. This is stated as follows:

Lemma 2.7
If $[\alpha_1] = [\alpha_2]$, and $[\beta_1] = [\beta_2]$, then

$$1 \quad [\alpha_1] \leqslant [\beta_1] \text{ iff } [\alpha_2] \leqslant [\beta_2]$$

and

$$2 \quad [\alpha_1]' = [\alpha_2]'.$$

This result is proved by noting appropriate results in the logic CBL. For example for 1, if $[\alpha_1] = [\alpha_2]$ and $[\beta_1] = [\beta_2]$, then $\vdash\alpha_1 \equiv \alpha_2, \vdash\beta_1 \equiv \beta_2$ and so if $\vdash\alpha_1 \supset \beta_1$, then $\vdash\alpha_2 \supset \beta_2$ also, as one can check by truth tables. A similar argument will establish 2.

So the relation $\leqslant$ and operation $'$ on CBL $+$ are well-defined, and one can investigate their properties. First, to justify the use of symbol $\leqslant$, one shows that

Lemma 2.8
CBL$+$ = $\langle$ CBL$+$, $\leqslant\rangle$ is a poset.

This follows from the properties of $\supset$ in CBL. For example, this relation is reflexive since $\vdash\alpha \supset \alpha$ for any wff α in CBL, and so $[\alpha] \leqslant [\alpha]$ in CBL $+$ by the definition of $\leqslant$. Similarly the properties of antisymmetry and transitivity follow from corresponding results in the logic, and so $\leqslant$ is a partial ordering, and CBL $+$ is a poset (see §2.2 above).

In fact, a stronger result about the structure of CBL $+$ can be established: that this poset is also a lattice. We show this by first noting that disjunction and conjunction are represented in CBL $+$ by lattice join and meet respectively, i.e.

Lemma 2.9

$$[\alpha] \bigvee [\beta] = [\alpha \vee \beta]$$
$$[a] \bigwedge [\beta] = [\alpha \cdot \beta].$$

The first part is proved by noting that $[\alpha \vee \beta]$ is an upper bound for the set $\{[\alpha], [\beta]\}$, since $\vdash\alpha \supset (\alpha \vee \beta)$ and $\vdash\beta \supset (\alpha \vee \beta)$, and so $[\alpha] \leqslant [\alpha \vee \beta]$ and $[\beta] \leqslant [\alpha \vee \beta]$. To show that it is in fact the lub of this set, suppose that there is some other upper bound, $[\gamma]$. Then $[\alpha] \leqslant [\gamma]$,

$[\beta] \leqslant [\gamma]$ by definition, and so $\vdash \alpha \supset \gamma$ and $\vdash \beta \supset \gamma$ by the definition of $\leqslant$. Hence $\models (\alpha \vee \beta) \supset \gamma$ as one can show by truth tables, and so $[\alpha \vee \beta] \leqslant [\gamma]$. Thus $[\alpha \vee \beta]$ is the lub $\{[\alpha], [\beta]\}$ as required. The second part of the lemma is proved by similar argument.

One can now establish the important result:

Lemma 2.10
CBL$+$ is a lattice.

This follows from the fact that for any wffs α, β in CBL, $\alpha \vee \beta, \alpha \cdot \beta$ are also well-defined wffs, and so for any elements $[\alpha]$, $[\beta]$ in CBL$+$ there are elements $[\alpha \vee \beta]$, $[\alpha \cdot \beta]$ also. By lemma 2.9 above, these elements are actually the lattice join and meet, respectively, of $\{[\alpha], [\beta]\}$. Thus by definition CBL$+$ is a lattice. The lemma could have been proved another way, by showing that the operations $\bigvee$ and $\bigwedge$ have the appropriate properties, L1–L4 of lattice operations (see lemma 2.1). These are easily established by appeal to the corresponding logical properties. For example, one shows that $\bigvee$ is idempotent, i.e. has property L1, by showing that $\models (\alpha \vee \alpha) \equiv \alpha$ holds in CBL and so $[\alpha] \bigvee [\alpha] = [\alpha]$ in CBL$+$. The other properties are established in a similar way. Thus the structure CBL$+ = \langle \text{CBL}+, \leqslant \rangle = \langle \text{CBL}+, \bigvee, \bigwedge \rangle$ is a lattice.

To understand fully the characteristics of $\leqslant$, or alternatively of $\bigvee$ and $\bigwedge$ in the structure representing CBL, we should note that not only is this structure a lattice, but it also satisfies the distributive identities (D1 and D2 of §2.2.1):

Lemma 2.11
CBL$+$ is a distributive lattice.

Again, this is proved by appeal to the properties of connectives in the logic. For example, $\models (\alpha \vee (\beta \cdot \gamma)) \equiv ((\alpha \vee \beta) \cdot (\alpha \vee \gamma))$ holds in CBL, so $([\alpha] \bigvee ([\beta] \bigwedge [\gamma])) = (([\alpha] \bigvee [\beta]) \bigwedge ([\alpha] \bigvee [\gamma]))$ in CBL$+$, which is to say that D1 is satisfied. D2 is proved by similar argument.

It is now interesting to turn to the operation $'$ in CBL$+$, which represents the connective $\neg$ in CBL. The properties of this operation are established as follows. First one can show that $'$ has two of the properties discussed in the last section:

Lemma 2.12
Operation $'$ satisfies C2 and C3 (the de Morgan and involution identities respectively).

Like the previous lemmas this follows from the properties of the connective which is represented. For example, since $\models \neg (\alpha \vee \beta) \equiv \neg \alpha \cdot \neg \beta)$, $([\alpha] \bigvee [\beta])' = [\alpha]' \bigwedge [\beta]'$ and so the first de Morgan identity (i.e. the first part of C2) is satisfied. The second is similarly established. To show C3, note that since $\models \neg\neg\alpha \equiv \alpha$, $([\alpha]')' = [\alpha]$ and so $'$ is an involution.

In fact, the operation $'$ is also a lattice complement in $CBL+$ and so satisfies condition C1 as well. To show this, one must first identify the universal bounds of this lattice:

Lemma 2.13

$$[1] = [(p \supset p)]$$
$$[0] = [\neg(p \supset p)]$$

This follows from the fact that for any wff α in CBL, $\models \alpha \supset (p \supset p)$ and $\models \neg(p \supset p) \supset \alpha$, as one can show by truth tables. Hence the equivalence class of $(p \supset p)$ and $\neg(p \supset p)$ are upper and lower bounds respectively of the lattice $CBL+$, by the definition of $\leqslant$.

Lemma 2.14
Operation $'$ satisfies C1 (i.e. is a lattice complement).

This is proved by noting that for any wff α in CBL, $\models(\alpha \lor \neg\alpha) \equiv (p \supset p)$ and $\models(\alpha \cdot \neg\alpha) \equiv \neg(p \supset p)$, and so $[\alpha] \lor [\alpha]' = [1]$ and $[\alpha] \land [\alpha]' = [0]$.

Since by lemmas 2.12 and 2.14 the operation $'$ has all the properties C1–C3 this is by definition an orthocomplement. Thus

Theorem 2.3
$CBL+$ is a Boolean Algebra (i.e. $CBL+$ is a distributive lattice in which every element has an orthocomplement).

This is the Representation Theorem for the classical logic CBL, and it summarises the essential characteristics of the connectives in this logic. Valid inference in CBL is represented by a partial ordering relation in a distributive lattice. Disjunction and conjunction have the properties of join and meet in such a lattice, and negation has all the properties of a lattice orthocomplement.

This discussion of CBL shows the significance of lattice theory to the study of logic. Logic is the study of systems of reasoning, and so is likely to be founded on some kind of inference relation. It is generally assumed that inference is at least a reflexive and transitive relation, as was the valid hook in CBL. Such a system will therefore be represented by a structure which is at least a poset. Furthermore, since propositional logics usually have rules of formation similar to those of CBL, with disjunction and conjunction introduced by similar definitions, it follows that (generally speaking) the representative structure is not only a poset, but a lattice. Note that even if disjunction and conjunction are regarded as fundamental to a logic, rather than implication, since these are usually assumed to have at least the defining properties of join and meet, on these grounds too one can suppose that the structure representing a logic will usually be at least a lattice.

The question of whether a logic must be represented by a *distributive* lattice has recently been much discussed. Clearly, one can easily generate abstract logics which do not have this property, but more interesting is the question of whether such a logic really represents any useful system of reasoning. Until the work of Birkhoff and von Neumann on 'quantum logic', it had generally been accepted that logical inference has all the properties of set inclusion. As these authors themselves remark (Hooker 1975 p 10), distribution is

the characteristic property of set combination

and so one might suppose that this property characterises logical inference also. However, these authors went on to argue that in quantum mechanics one actually employs a system of reasoning which is not distributive. Much work has since been done investigating lattices which do not have this property. Birkhoff and von Neumann suggested that a weaker property, called 'modularity', may characterise all lattices representing inferential systems. Since then there have been many other suggestions.

The arguments supporting the 'quantum logicians'' claims will be thoroughly analysed later in this book. We can, however, remark that distribution does appear to be essential to our ordinary understanding of inference. Any non-distributive lattice would represent a system of reasoning radically foreign to everyday understanding.

Of course it is not only the failure of distribution which might prevent a logic from being represented by a Boolean algebra. Instead, there may be no lattice orthocomplement in the representative algebra. It has been suggested often in the history of logic that the classical 'Law of Excluded Middle' might fail to be universally valid. This 'law', represented by the tautology $\alpha \lor \neg\alpha$ in CBL, asserts that every wff in the logic must take one of the two truth-values in every valuation. It gives rise in the representative algebra, to the property of lattice complement for the operation representing negation. In the next chapter we shall investigate a classical but non-bivalent logic, in which the 'law' of Excluded Middle fails. We shall see that this system is represented by a lattice which is distributive but which has no lattice complement. Instead, negation is represented by a relative orthocomplement. Since all connectives are represented by operations with essentially the same properties as those representing connectives of CBL, one can argue that although it is not represented by a Boolean algebra, this logic is essentially classical.

3 The Propositional Logic of Mechanics

3.1 The logic of mechanical theories

We now discuss the propositional logic which is used to express the description of mechanical theories. We regard a theory T as an individual system of classical or quantum mechanics used to describe some individual reality. According to §2.1, we shall determine the logic used by a theory firstly when we identify its simple propositions, secondly when we know how these are combined to form complex descriptions of reality and lastly when we know how valuations and logical connectives are defined. Each of these points will be discussed in turn, so that any theory T will be associated with a logic LM_T, which expresses its descriptions of reality. It will also be possible to generate an abstract propositional 'logic of mechanics' LM, which has the characteristics common to the logic of any mechanical theory.

3.1.1 Simple propositions

To identify the simple propositions in mechanical theories, recall the discussion of classical and quantum mechanics in §1.1. Recall that a classical theory can be represented in terms of an underlying phase space, parametrised by position and momentum coordinates. Observables, such as position, mass or energy, are represented by real-valued functions of the phase variables. A classical state is represented by a point in phase space, and so corresponds to an assignment of values to all the observables. The value assigned to each observable is understood as the outcome of a measurement of the observable in this state. For example, if value 3 is assigned to the observable mass in a classical state S, then we say that mass has value 3, or the measurement outcome of mass is 3 in state S†.

---

† These two formulations are assumed to be equivalent, because of a view of measurement which is discussed in chapter 4. In this book it is only the logical properties of measurement which are considered, and so any distinction between these propositions is irrelevant.

Recall that quantum theories are not represented by means of a phase space, but instead by means of a vector space of states. States of a quantum theory are represented by special vectors in the space, and observables by special operators on the vectors. The spectrum of any observable operator contains only real eigenvalues, which are understood as the allowable measurement outcomes of the observable which is represented. For example, if ψ is a state vector which is the eigenvector corresponding to eigenvalue 2 in the spectrum of observable operator Q, representing position, then we say that in state ψ position has value 2 in this state.

The simple propositions in either theory thus appear to have similar form. Recall from §2.1.1 that a proposition is an expression which can be true or false, and that a simple proposition is one that is not further analysed by the logic into constituent propositions. Expressions such as those above: the measurement outcome of mass is 3 or position has value 2 meet both requirements. These are the simple, elementary descriptions used by mechanical theories.

To give a formal definition of the simple propositions of any mechanical theory T, consider the primitive concepts from which they are generated. Any theory T of mechanics uses a set M_T of *magnitudes*, or *observables*. These were characterised earlier in both classical and quantum theories. In classical theories they are represented by real-valued functions of the phase variables, and in quantum theories by observable operators in Hilbert Space. Corresponding to every magnitude m in the set M_T of a theory, is a value-set V_m, which contains all its allowable values, or possible outcomes of measurements. The set V_m is thus a subset of real numbers. The value-sets of observables in classical and quantum theories were characterised earlier as the values of the real-valued functions, or as elements in the spectrum of the observable operators, respectively.

A simple proposition of any mechanical theory T may thus be represented formally as follows:

Definition

A *simple proposition* of mechanical theory T, is an ordered pair (m, Δ), where $m \in M_T$, and $\Delta \subseteq V_m$ is a Borel subset of this value-set.

Informally, we understand any proposition (m, Δ) as the statement that magnitude m has a value in Δ. Propositions of this form can be arbitrarily precise, so that for example (m, r), m has value r is considered a special case of the definition, for r a real number in V_m. Such propositions (m, r) are called *atomic*. The requirement that Δ be a Borel subset ensures not only that propositions can be made arbitrarily precise in this way, but also that set operations among the subset of V_m are well-defined. This is important as it allows corresponding relations and operations to be defined among propositions, as we shall discuss below.

The set of all simple propositions of a theory T will be known as its elementary set E_T.

Definition
The *elementary set* E_T of theory T is its set of simple propositions, i.e.
$E_T = \{(m, \Delta): m \in M_T$ and Δ a Borel subset of $V_m\}$.

It follows from the definition that E_T will be made up of subsets of propositions, each associated with a particular magnitude in M_T. It is useful to define

Definition
p is an *m-proposition* of T if $p \in E_T$ and $p = (m, \Delta)$, for some Δ. The *m-subset* of E_T is the set P_m where

$$P_m = \{p: p \text{ is an } m\text{-proposition of } T\}.$$

The relation of set-inclusion among subsets of each value-set V_m and the operation of set-complement in each value-set generates corresponding relations in each *m*-subset of simple propositions. These are defined as

Definition
(Inference) $(m, \Delta_1) \leqslant (m, \Delta_2)$ iff $\Delta_1 \subseteq \Delta_2$
(Relative complement) $(m, \Delta)^\perp = (m, V_m - \Delta)$.

(The symbol $^\perp$ indicates a lattice orthocomplement: see below.) It follows from the properties of set-inclusion that inference is indeed a partial ordering of propositions and so the elementary set E_T of any theory T is at least a poset.

Lemma 3.1
$E_T = \langle E_T, \leqslant \rangle$ is a poset (where $\leqslant$ is the inference relation defined above).

However, each *m*-subset of the elementary system E_T is not only a poset but a Boolean algebra, for within each P_m the set operations generate corresponding Boolean operations among propositions. For example, set-union and set-intersection generate the following operations in any P_m:

Definition
(join) $(m, \Delta_1) \vee (m, \Delta_2) = (m, \Delta_1 \cup \Delta_2)$
(meet) $(m, \Delta_1) \wedge (m, \Delta_2) = (m, \Delta_1 \cap \Delta_2)$.

It follows from the properties of set operations that each P_m is a distributive lattice with respect to these operations. To show that the operation of relative complement defined above is a lattice orthocomplement in each *m*-subset of E_T, one must first identify the upper and lower bounds of these subsystems. These are in fact $1_m = (m, V_m)$, and

$0_m = (m, \emptyset)$ respectively where $\emptyset$ is the empty set, as one can easily check. It follows simply from the properties of set-complement in V_m that operation $\perp$ above is a lattice orthocomplement in P_m. Thus

Lemma 3.2
For each m in M_T, P_m is a Boolean algebra.

Hence, from lemma 3.1 also, one concludes

Theorem 3.1
For any theory T, the elementary set E_T is a union of Boolean algebras.

Note that E_T is still only a poset, not a lattice, since arbitrary join and meet, for example between propositions from different m-subsets, are not defined.

   In fact, the elementary set of a theory will generally have additional properties to those indicated in theorem 3.1 above. There will generally be fundamental laws of a theory which specify relations between magnitudes in M_T, and hence generate additional relations and operations among simple propositions in E_T. Newton's second law in classical mechanics is an example, expressing a functional relation between magnitudes of any classical theory. The law asserts that force = mass × acceleration. This relation among magnitudes generates corresponding relations among the simple propositions of a theory of classical mechanics. Inference relations are generated between propositions from the mass, force and acceleration subsets. Thus any classical theory T will have these additional relations in its elementary set E_T of simple propositions. Similarly there are other classical laws generating other propositional relations. Quantum theories will also have laws which generate additional elementary structure.

   However, in this discussion of logic we are concerned only with the structure common to any elementary system, given in theorem 3.1 above. Any system E_T has an inference relation which is a partial ordering of propositions, and set relations on the value-set of each magnitude generate corresponding relations among the m-subsets of propositions, so that each m-subset is in fact a Boolean algebra of simple propositions.

3.1.2 Complex propositions

So far we have seen that any theory T will have a set E_T of simple propositions, each of which is represented by an ordered pair (m, Δ) for some magnitude m in M_T and Δ, a Borel subset of V_m. In the formal logic of this theory, these simple propositions are represented by the propositional variables p, q, r, ... Complex formulae in the logic, representing complex propositions, can then be formed from these by means of logical connectives.

We saw in §2.1 that a small number of connectives are fundamental to a propositional logic. Rules of formation determine how they are used to form well-formed expressions. They are defined by valuation rules which specify how the truth-value of a complex proposition depends on that of its constituent variables. For an abstract system such as CBL, generated simply from a set of propositional variables, the choice of connectives may seem an arbitrary matter. We are bound only by our desire to express a particular system of reasoning. In discussing the logic of a theory, however, the case is somewhat different. Here we are also generating a logic from a set of propositional variables. However, we know in this case that the variables represent propositions of a particular form, which are interrelated in a particular way, described above. In considering the logic of a mechanical theory we must take these interrelations into account.

There are in fact two ways in which this structure influences the logic appropriate to mechanical theories. Simple valuations, the truth-value assignments to simple propositions of a theory, must be consistent with the relations among propositions in the system E_T. In addition, the logic of complex propositions must include connectives which are capable of expressing these relations.

For this latter reason, we include at least a unary connective $\neg$ in the logic of mechanical theories, which is to represent the relative complement operation $\perp$ among simple propositions. We also include a binary connective $\supset$ to express propositional inference. Thus, as with the logic CBL discussed in the last chapter, we have two primitive connectives $\neg$ and $\supset$ in the logic of mechanical theories. In fact, the alphabet and rules of formation for the logic LM_T of a mechanical theory T will be just those of CBL.

Alphabet of LM_T

p, q, r, . . .   variables representing simple propositions in E_T
$\neg$, $\supset$ a binary and a unary connective respectively
(,)   brackets

Rules of Formation for LM_T

1 Any propositional variable p is a wff of LM_T.
2 If α, β are wffs, then so are $\neg\alpha$ and $\alpha \supset \beta$.
3 Only those strings generated from rules 1 and 2 are wffs. Normal conventions of bracketing apply.

These rules tell us how the wffs in the logic of a mechanical theory are generated. To give meaning to the wffs, we must consider the nature of valuations in such a logic.

3.1.3 Valuations

Recall that a valuation is an assignment of truth-values to simple propositions, and that in any mechanical theory simple propositions are elementary descriptions of reality. It follows that valuations of a theory correspond to descriptions of a system, since by assigning truth-values to elementary propositions we do express a description of reality. For example, if we hold that 'mass has value 3' is true of a physical system, while perhaps 'position has value 2' is false, we do give some kind of description of this system.

However, not every truth-value assignment to members of E_T will count as a description given by T, and so not every one will count as a valuation. For one thing, truth-value assignments must be consistent if they are to describe reality. If we say that 'mass has value 3' is true, then we cannot for example hold that 'mass has value 5' is true also. In fact, we require at least that the descriptions given by T are consistent with the two fundamental relations on E_T discussed above. In addition, they must be consistent with the laws of T.

These conditions of consistency can be expressed formally. In the case of inference, we require that whenever $p \leqslant q$ holds on E_T, a valuation must find q true when p is true, and must find p false if q is false, for as remarked in the earlier discussion of propositional inference (§2.1.2), we do not allow an inference to lead us from truth to falsity. A true antecedent will yield a true consequence if the inference holds, and a false consequence can only mean that the antecedent is also false. These are generally accepted as minimal assumptions about the nature of propositional inference. As far as the relative complement operation is concerned, consistency requires that a valuation assigns opposite truth-values to relative complements. If $p = (m, \Delta)$ is true, then we understand that the value of m is in Δ, and so $p^{\perp} = (m, V_m - \Delta)$, the proposition that the value of m is outside Δ is in this case judged to be false. Similarly, when $p^{\perp}$ is true, its relative complement p is false. Only truth-assignments which meet these consistency conditions can be regarded as valuations of the theory, representing its descriptions of reality.

In addition to these requirements of consistency, we also require that valuations be *coherent*, in the sense that they assign at most a single truth-value to every proposition. Clearly, any truth-value assignment which finds a single proposition both true and false cannot express any coherent description. It follows from this requirement that a valuation can be defined as a function from the system E_T to the two truth-values, for by definition a function maps each element to at most one member of its range.

To define valuations formally as functions we need to consider the nature of the co-domain. In particular, we must consider whether truth-value 'gaps' are allowed, or whether every proposition must always be mapped to one of the two truth-values. Since valuations are to represent the descriptions given by a theory, we must consider whether descriptions are expressed when truth-value gaps occur. I think one must conclude that they are, for we do describe reality, even if we conclude no more than that 'mass has value 3' is a true elementary description. In such a description we know that this proposition p is true, and so by the requirement of consistency we know that other mass propositions will also be true and others false. This constitutes a description of reality even when the truth-values of other propositions remain unknown. In classical mechanics, for example, we can describe a system even if we are unable to determine the precise values of *every* magnitude.

Thus we conclude that valuations of mechanical theories need not be bivalent truth-value assignments. They need not map every simple proposition in E_T to one of the two truth-values. We should remark here that according to our analysis there will actually be theories in which valuations could not be bivalent, for in general one cannot assume that a function exists which maps every element of a poset E_T, in a structure-preserving way, onto the two-element set of truth-values[†].

Since valuations are not required to map every simple proposition onto the two truth-values, we can represent them by functions onto a three-member set $\{t, f, u\}$. Whenever a truth-value gap occurs we consider the proposition 'undecided' and assign it to the third element u. Those propositions mapped to t are regarded as *true* in the valuation, those assigned to f are *false*, while those assigned to u are *undecided* propositions, which have no truth-value in this valuation. The logic founded on these valuations will be a three-valued logic since there are these three possible relations between a proposition and any valuation. Note, however, that there are only two truth-values. To call a proposition undecided is just to say that it lacks a truth-value.

Thus we define

Definition

A (*simple*) *valuation of theory* T is a function h from the elementary set E_T to a three-element set $\{t, f, u\}$, $h: \{p, q, r, \ldots\} \rightarrow \{t, f, u\}$ such that the following hold:

    1 $h(p) = t$ iff $h(p^{\perp}) = f$
    2 If $p \leqslant q$ in E_T, then if $h(p) = t$, $h(q) = t$, and if $h(q) = f$, $h(p) = f$
    3 h conforms to the laws of T

Condition 1 ensures that h is consistent with the operation $^{\perp}$ on E_T, as

† The algebra containing only the elements $\{0, 1\}$ is called Z_2. Not every poset can be mapped onto this two-element algebra in a structure-preserving way.

discussed above. Note that by the involution property of $^\perp$ in each *m*-subset P_m, $(p^\perp)^\perp = p$ for any p and so the converse of condition 1 also holds, and $h(p^\perp) = t$ iff $h(p) = f$. These propositions always take opposite truth-values. Condition 2 requires that *h* be consistent with the relation $\leqslant$ among simple propositions, in accordance with earlier discussion. Condition 3 generates different formal constraints according to the particular theory which is considered.

When we consider a particular logic LM_T which expresses the descriptions of a particular theory T, then we must take condition 3 of this definition into account, for the valuations of T must be consistent with the laws of this theory. However, when we wish to consider the abstract logical system LM which has only the characteristics common to the logic of any mechanical theory T, we can ignore condition 3. This abstract logic will be discussed in §3.1.5 below.

3.1.4 Logical connectives

Having defined the simple valuations of a theory, one can give valuation rules which define the logical connectives. Recall that these rules extend any simple valuation assigning truth-values to simple propositions to a complex valuation which assigns truth-values to all wffs in the logic of a theory. We saw above that two connectives $\supset$ and $\neg$ are taken as fundamental in any logic LM_T. These are required to express the relations $\leqslant$ and $^\perp$ among simple propositions and so must have some appropriate properties. In fact these properties are given in conditions 1 and 2 of the definition of a valuation. In addition, we require that $\leqslant$ generates an inference relation in the sense discussed earlier (in §2.3). These requirements lead us to conclude that the following connectives are appropriate to the logic of a theory.

Table. Rules defining $\neg$ and $\supset$ in LM_T.

$\neg$	α		α	$\supset$	β
f	t		t	t	t
t	f		t	f	f
u	u		t	u	u
			f	t	t
			f	t	f
			f	t	u
			u	t	t
			u	u	f
			u	t	u

The negation of α, $\neg\alpha$, takes the opposite truth-value to α and so meets the requirement of condition 1 in the definition of a valuation (see also lemma 3.3 below). The connective is thus suitable to express the operation $^{\perp}$ on E_T. Notice that this connective coincides with negation in classical bivalent logic, where only the bivalent valuations are considered—that is, if we ignore all occurrences of u under wff variables in the truth table for $\neg$ above, it coincides with the table defining $\neg$ in CBL (§2.1).

Similarly, the hook defined above also coincides with $\supset$ in CBL, where only bivalent valuations are considered. Furthermore, condition 2 of the definition of a valuation is respected, making $\supset$ suitable to express the inference relation $\leqslant$ on E_T (see also lemma 3.4 below). Note that this connective $\supset$ has the essential property of the bivalent hook, in that it can never lead us from true antecedents to a false conclusion. A wff $\alpha \supset \beta$ is always false if α is true but β false. We see in rows 3 and 6 that where α is true but β undecided, the wff itself also lacks a truth-value, and similarly where α is undecided but β is false, the wff is undecided. Row 7 shows that in all cases where the consequent is true, so too is the implication. According to the last row of the table, when both constituents are undecided the wff is true. This condition is required so that $\leqslant$ can represent an inference relation, in the sense discussed above (§2.3). This condition ensures that $\alpha \supset \alpha$ is a tautology of the logic, and so $\supset$ generates a reflexive relation.

Some useful connectives can be introduced into the logic of any theory by the following definitions:

Abbreviations for LM$_T$

1	(Disjunction)	$\alpha \vee \beta = $ df $(\alpha \supset \beta) \supset \beta$
2	(Conjunction)	$\alpha \cdot \beta = $ df $\neg(\neg\alpha \vee \neg\beta)$
3	(Biconditional)	$\alpha \equiv \beta = $ df $(\alpha \supset \beta) \cdot (\beta \supset \alpha)$
4	(Denial)	$\sim\alpha = $ df $\alpha \supset \neg\alpha$
5	(True)	$T\alpha = $ df $\neg(\sim\alpha)$
6	(False)	$F\alpha = $ df $\neg(\neg\alpha \supset \alpha)$
7	(Undecided)	$U\alpha = $ df $\neg(T\alpha \vee F\alpha)$

The first three definitions are familiar from the earlier discussion of CBL. The following truth tables characterise these connectives and one can see from these that in fact they do coincide with classical bivalent connectives where bivalent valuations alone are considered.

Table. Truth values for connectives in LM_T.

α	β	~α	Tα	Fα	Uα	α ∨ β	α · β	α ≡ β
t	t	f	t	f	f	t	t	t
t	f					t	f	f
t	u					t	u	u
f	t	t	f	t	f	t	f	f
f	f					f	f	t
f	u					u	f	u
u	t	t	f	f	t	t	u	u
u	f					u	f	u
u	u					u	u	t

Note that the connective $\sim$ is called denial since $\sim\alpha$ is true whenever α fails to be true. It is not, like negation, only true when α is actually false. This connective is also a generalisation of classical bivalent negation, since it too coincides with this negation when only bivalent valuations are considered. However, it is not chosen to express the relation $^\perp$ among simple propositions, as it does not meet condition 1 in the definition of a valuation. α and $\sim\alpha$ do not always take opposite truth-values, since $\sim\alpha$ will be true when α is undecided. Denial is used to assert that a wff fails to be true, while negation can make the stronger claim, that a wff is actually false. In a bivalent logic this distinction cannot be made. The connectives T, F, and U allow us to express in LM_T the relation between any wff and any valuation.

Disjunction and conjunction also coincide with the classical bivalent connectives when only bivalent valuations are considered, as one can easily check from the tables above. Both share the essential properties of these connectives. Thus a disjunction is true only when at least one constituent is true, while a conjunction is true only when both are true. The cases involving truth-value gaps are easily understood. The biconditional connective $\equiv$ is also a generalisation of the connective in CBL and shares its essential property of being true only when both constituents have the same value in a valuation. Again we will find this connective useful in discussing the algebraic structure of the logic.

3.1.5 Some properties of logic LM

We now examine the logic which is shared by mechanical theories. To do this we generate an abstract system LM from the definitions and rules above which are common to the logic of any mechanical theory. For instance, the Alphabet of LM contains the same symbols as the alphabet of LM_T, i.e. propositional variables, connectives symbols and brackets. It is also assumed that the set $E = \{p, q, r, \ldots\}$ of propositional variables has the structure common to each E_T, i.e. $E = \langle E, \leqslant, {}^\perp \rangle$ is a union of

Boolean algebras. The Rules of Formation for LM are the same as for each LM_T, so that wffs in all these systems are the same. A Valuation of LM is a function from set E to the set $\{\text{t, f, u}\}$, which is consistent with $\leqslant$ and $\perp$, i.e. which meets conditions 1 and 2 in the earlier definition of a valuation of T. The Valuation Rules defining the connectives of LM are just those given above for LM_T, and the same Abbreviations are used to introduce additional connectives.

Thus LM has the structure which is common to all the individual logics LM_T. The wffs which are valid in LM will also be valid in any logic LM_T. The converse, however, is not the case, for wffs may be true in every valuation of T, just because of its own peculiar laws and the consistency conditions these impose (by condition 3 of the definition of a valuation). It is in fact useful to distinguish, in any LM_T, between the wffs which are logical truths in the sense of being valid in LM, and those which one might call 'theoretical' truths.

Definition
α is a *logical truth* in LM_T, $\models \alpha$, if α is valid in LM (i.e. if $\models_h \alpha$ for every valuation h of LM).
α is a *theoretical truth* in LM_T, $\text{T} \models \alpha$, if α is valid in LM_T (i.e. if $\models_h \alpha$ for every valuation h of T).

Clearly, any logical truth is also theoretically true, but the converse does not hold. The logical truths are valid in any logic LM_T, while the theoretical trueths may be valid only for some particular theory T. It is the logical truths of LM, expressing the common structure of the individual systems LM_T which are of interest here, and which will now be briefly investigated before the algebraic discussion in §3.2 below.

In discussing the properties of LM it is useful once more to introduce valuation symbols to express the relations between wffs and valuations. As in chapter 2 we use the symbols $\models_h \alpha$, $\dashv_h \alpha$ to indicate that α is true in h or false in h respectively, with corresponding symbols without subscripts to represent validity or logical falsity (see §2.1). In addition, we now require the following symbols to represent the other relations:

Definition

$\perp\!\!\!\perp_h \alpha$	iff $h(\alpha) = \text{u}$
$\perp\!\!\!\perp \alpha$	iff $h(\alpha) = \text{u}$ for every valuation h of LM

In the first case α is *undecided in h*, i.e. it lacks a truth-value in this valuation, and in the second it is *logically undecided*. As before, a stroke through the relation symbol indicates that it fails to hold, so for example $\not\perp\!\!\!\perp_h \alpha$ means that $h(\alpha) \neq \text{u}$.

First, let us show that the choice of connectives in LM is indeed suitable to express the relations among simple propositions. Recall that the set E of propositional variables is assumed to be a poset which is a union of Boolean algebras, $E = \langle E, \leqslant, {}^{\perp} \rangle$. We now show that

Lemma 3.3
$$\models (p^{\perp}) \equiv (\neg p).$$

This is established by noting that $h(p^{\perp}) = t$ iff $h(p) = f$, iff $h(\neg p) = t$, by the definitions of h and $\neg$. Similarly $h(p^{\perp}) = f$ iff $h(p) = t$, iff $h(\neg p) = f$, by the definitions of h and $\neg$. It therefore follows also that $h(p^{\perp}) = u$ iff $h(\neg p) = u$, and so the lemma is proved by the definition of $\equiv$.

Thus the connective $\neg$ is appropriate to express the operation ${}^{\perp}$ among simple propositions. A similar result holds for $\supset$, as we now show

Lemma 3.4
If $p \leqslant q$, then $\models p \supset q$

This follows from condition 2 on the definition of a valuation, and can be established by examining the truth table defining $\supset$.

One can now show that some important tautologies of classical bivalent logic are also valid in LM:

Lemma 3.5
$$\models \alpha \supset \alpha$$
$$\models ((\alpha \supset \beta) \cdot (\beta \supset \gamma)) \supset (\alpha \supset \gamma)$$
$$\models \neg(\neg \alpha) \equiv \alpha$$

Each of these results can be established by a truth table. There are, however, other tautologies of CBL which fail to be valid in LM. Most important of these are perhaps the classical 'Laws' of Excluded Middle and Non-Contradiction, which in a sense characterise any bivalent system:

Lemma 3.6
$$\not\models \alpha \vee \neg \alpha$$
$$\not\models \neg(\alpha \cdot \neg \alpha)$$

Both results are proved by considering a valuation h in which wff α is undecided. If $h(\alpha) = u$, then $h(\neg \alpha) = u$, and so $\perp\!\!\!\perp_h \alpha \vee \neg \alpha$ and also $\perp\!\!\!\perp_h \alpha \cdot \neg \alpha$. From the first part of this lemma one can derive the following result:

Lemma 3.7
$$\not\models \beta \equiv ((\beta \cdot \alpha) \vee (\beta \cdot \neg \alpha)).$$

Since $(\alpha \vee \neg \alpha)$ is not valid, $\beta \cdot (\alpha \vee \neg \alpha)$ is not equivalent to β, and so by distribution the lemma holds.

We saw in chapter 2 that a propositional logic can be represented axiomatically if it is both sound and complete. (See §2.1.4.) The logic LM has been studied before for different reasons, most notably by the Polish logician Lukasiewicz whose name is sometimes given to the system†. M Wajsberg has shown that this logic can be axiomatised (see McCall 1967). As for CBL, one takes Modus Ponens and Uniform Substitution as the Rules of Inference. For LM the axioms are as follows.

Axioms for LM

1. $\alpha \supset (\beta \supset \alpha)$
2. $(\alpha \supset \beta) \supset ((\beta \supset \gamma) \supset (\alpha \supset \gamma))$
3. $(\neg\alpha \supset \neg\beta) \supset (\beta \supset \alpha)$
4. $((\alpha \supset \neg\alpha) \supset \alpha) \supset \alpha$

Further discussion and references to this axiomatisation can be found for example in Rescher (1969).

3.1.6 Summing up

We have now applied the methods of modern logic, described in §2.1 to mechanical theories. From the earlier discussion in chapter 1 of classical and quantum mechanics, it was assumed that any mechanical theory T specifies a set M_T of magnitudes, each of which is associated with a set of real numbers, its value-set. A theory also specifies laws, which establish relations among its magnitudes.

The simple propositions which express a theory's elementary descriptions of reality, are of the form (m, Δ), with m a magnitude of the theory and Δ a Borel subset of its value-set. (m, Δ) can be understood as the statement that The value of m is in Δ. Set relations among the subsets of each value-set generate corresponding relations among the simple propositions, so that each set P_m of m-propositions has the structure of a Boolean algebra. It follows that the elementary set E_T of simple propositions used by theory T is at least a union of Boolean algebras—each algebra is associated with one of its magnitudes. In general there will be other relations among the propositions, generated by the laws peculiar to that theory.

To generate a logic LM_T from the propositions of a theory, valuations are defined as truth-value assignments which preserve the structure of E_T. In particular, they must be consistent with relations $\leqslant$ and $^\perp$ among simple propositions. Logical connectives $\supset$ and $\neg$ are chosen with properties appropriate to express these relations and so a propositional logic is generated. A three-valued logic LM_T which allows truth-value gaps is appropriate to express the descriptions of any mechanical theory T.

In the last section we examine the logic LM, which has the structure common to any LM_T. We saw that several results characterising classical

† Lukasiewicz was probably the first to develop this logic. See, for example, Borokowski (1970) or McCall (1967). The system LM is sometimes called L_3 after Lukasiewicz.

bivalent logic also hold in LM, although naturally the 'Laws' of Excluded Middle and Non-Contradiction do not hold in this system. In the next section a more thorough analysis of the logic is given, using the algebraic methods described in the last chapter.

3.2 The algebraic structure of the logic

3.2.1 *The structure of LM*

To investigate the structure of logic LM and to compare it with classical bivalent logic, we shall turn to the algebraic methods discussed in §2.2 above. We construct an algebra to represent the logic and then investigate its properties. In this way we shall see that the connectives of LM are essentially classical, except that negation is slightly generalised. We can understand this logic as having a 'context' for each wff, or having a relevance relation among its formulae.

The algebra representing logic LM will be constructed following the usual Lindenbaum–Tarski method, described in chapter 2. Wffs of the logic are to be represented by classes of logically equivalent formulae, and connectives are represented by corresponding relations and operations among these classes. As in CBL, we can use the connective $\equiv$ to express logical equivalence, for as in the classical case wff $\alpha \equiv \beta$ is true only when both constituents share the same value in a valuation. Thus we can define a relation R among wffs by

$$\alpha \, R \, \beta = \models \alpha \equiv \beta.$$

This relation is reflexive, symmetric and transitive, as one can easily check, and so it generates the following equivalence classes of wffs:

$$[\alpha] = \{\beta : \models \alpha \equiv \beta\}.$$

The set of all such equivalence classes is denoted by

$$\text{LM}+ = \{[\alpha] : \alpha \text{ is a wff of LM}\}.$$

This is the carrier set of the algebra representing LM.

As in the earlier case of CBL we wish to represent $\supset$ and $\urcorner$ in this algebra, and as in this case we generate a relation to represent valid implications in the logic, and an operation to represent negation.

Definition

$$[\alpha] \leqslant [\beta] \qquad \text{iff } \models \alpha \supset \beta$$
$$[\alpha]' = [\urcorner\alpha].$$

It can be shown that these are well-defined by arguments analogous to those used in the case of CBL in §2.2 above. Similarly, one can show that $\leqslant$ is indeed a partial ordering of equivalence classes, and so:

Lemma 3.8
$$LM+ = \langle \text{LM}+, \leqslant \rangle \text{ is a poset.}$$

By arguments similar to those used earlier one can also show that disjunction and conjunction in LM correspond to the operations of lattice join and meet in LM+. That is,

Lemma 3.9
$$[\alpha]\vee[\beta] = [\alpha \vee \beta]$$
$$[\alpha]\wedge[\beta] = [\alpha \cdot \beta]$$

It follows that $LM+$ is a lattice, for the logical connectives are defined for any pair of wffs, and so operations $\vee$ and $\wedge$ are defined for any pair of equivalence classes. In fact, it follows from the properties of the connectives in LM that a stronger result holds:

Lemma 3.10
$$LM+ = \langle \text{LM}+, \leqslant \rangle = \langle \text{LM}+, \vee, \wedge \rangle \text{ is a distributive lattice.}$$

This tells us that the inferential structure of LM is essentially classical. The connectives $\supset$, $\vee$ and $\cdot$ of CBL also generate operations of a distributive lattice, as we saw in §2.2. These connectives are essentially the same in both LM and CBL.

Negation does appear to be different in these logics. First we give a result which shows their similarity in either system.

Lemma 3.11
Operation ′ on $LM+$ satisfies the de Morgan identities and is an involution (that is, ′ satisfies C2 and C3 of §2.2).

These results follow from the tautologies of LM. We saw for example, in §3.1.5 above, lemma 3.5, that $\models \neg(\neg\alpha)\equiv\alpha$, and so it follows that $([\alpha]')'=([\alpha])$ in $LM+$, and so by definition ′ is an involution (i.e. condition C2 holds). Similarly one shows that since $\models\neg(\alpha \vee \beta) \equiv (\neg\alpha \cdot \neg\beta)$ in LM, $([\alpha]\vee[\beta])' = [\alpha]'\wedge[\beta]'$ in $LM+$ and so one of the de Morgan identities is satisfied. The other is proved likewise.

However, it is interesting to note that ′ is *not* a lattice complement, as it is in $CBL+$. Condition C1 fails. To establish this result one must first identify the universal bounds of lattice $LM+$:

Lemma 3.12
$[1] = [p \supset p]$ and $[0] = [\neg(p \supset p)]$ are the universal upper and universal lower bounds of lattice $LM+$ respectively.

As in the case of CBL this follows from the fact that $\models\alpha \supset (p \supset p)$ and $\models\neg(p \supset p) \supset \alpha$ for any α. To show that ′ is not a lattice complement in $LM+$ we show for arbitrary α,

Lemma 3.13

$$[\alpha] \bigvee [\alpha]' \neq [1]$$
$$[\alpha] \bigwedge [\alpha]' \neq [0].$$

These identities fail because the corresponding wffs are not logically true and logically false. For instance, since it can be the case that $\underset{h}{\perp}\alpha \vee \neg\alpha$, $\nvDash (\alpha \vee \neg\alpha) \supset (p \supset p)$, and so the first identity fails. Similarly because $\underset{h}{\perp}\alpha \cdot \neg\alpha$ for some h, $\nvDash\alpha \cdot \neg\alpha$, and so this wff is not logically equivalent to $\neg (p \supset p)$, and the second identity fails. Thus the fact that $'$ is not a lattice complement stems directly from the non-bivalence of LM, and the resulting failure in this logic of the 'Laws' of Excluded Middle and Non-Contradition (see lemma 3.3).

At first it may seem that the failure of $'$ to satisfy C1, the lattice complement property, indicates that negation in this logic is radically non-classical. However, this is not the case. It will now be shown that negation in LM has all the properties essential to the classical connective, but only within a restricted subsystem of the logic. One can understand LM as a classical logic in which every wff has a context, or scope, or alternatively as a logic which has a *relevance* relation among its wffs.

The following definition will make this claim precise:

Definition
The *context* (or *scope*) of wff α in LM is the set

$$C_\alpha = \{\beta : \vDash (\alpha \cdot \neg\alpha) \supset \beta \text{ and } \vDash \beta \supset (\alpha \vee \neg\alpha)\}.$$

This is the set of wffs which are validly implied by the conjunction of α with its negation, and which validly imply the disjunction of α with its negation. From this set one can define a relevance relation Re:

Definition
β Re α in LM, iff $\beta \in C_\alpha$

Wff β is relevant to α if it is contained in its context.

The significance of the context of a wff is appreciated if we turn again to the representative algebra $LM+$, and in particular the subset $C_\alpha+$ which represents the context of a wff. That is

$$C_\alpha+ = \{[\beta] : \beta \in C_\alpha\}.$$

It follows from the definition of logical equivalence that this set is well-defined. In fact, one can show that the set is a subalgebra of $LM+$, i.e. it is closed with respect to the operations.

Lemma 3.14
$C_\alpha+$ is a subalgebra of $LM+$.

To prove this, note first that $C_\alpha+$ is non-empty, since it contains at least the element $[\alpha]$. To show closure with respect to join, suppose $[\beta]$, $[\gamma]$ are in $C_\alpha +$. Then $\models (\alpha \cdot \neg\alpha) \supset \beta, \gamma$ and $\models \beta, \gamma \supset (\alpha \vee \neg\alpha)$, by the definition of C_α. Thus $\models (\alpha \cdot \neg\alpha) \supset (\beta \vee \gamma)$ and $\models (\beta \vee \gamma) \supset (\alpha \vee \neg\alpha)$ as one can establish by truth tables. It follows that $(\beta \vee \gamma)$ is in C_α by definition, and so $[\beta] \bigvee [\gamma]$ is in $C_\alpha +$, by lemma 3.9 above. Similarly, one proves closure with respect to meet. To show closure w.r.t.', suppose $[\beta]$ is in $C_\alpha +$. Then $\models (\alpha \cdot \neg\alpha) \supset \beta$ and $\models \beta \supset (\alpha \vee \neg\alpha)$ and so $\models (\alpha \cdot \neg\alpha) \supset \neg\beta$ and $\models \neg\beta \supset (\alpha \vee \neg\alpha)$, as one can show by truth tables. Thus $\neg\beta$ is in C_α, and so $[\beta]'$ is in $C_\alpha +$. It follows that if wffs β, γ are relevant to a wff α, so too are all complex wffs formed from these.

The importance of the context of a wff is revealed in the following result:

Lemma 3.15
$[\alpha]'$ is the complement of $[\alpha]$ relative to $C_\alpha +$.

This result follows simply from the definition of C_α. By its construction, $[\alpha \cdot \neg\alpha]$ and $[\alpha \vee \neg\alpha]$ are the lower and upper bounds of $C_\alpha +$ respectively, and so it follows that lemma 3.15 holds. Thus within the subsystem $C_\alpha +$ of $LM +$, operation ' has the property C1: it is a lattice complement. Since operation ' has already been shown to have properties C2 and C3 (by lemma 3.11 above), it follows that within $C_\alpha +$ it has all the classical properties of an orthocomplement. This gives us the following representation theorem for the logic LM:

Theorem 3.2
$LM +$ is a distributive lattice, in which every element has an orthocomplement relative to some sublattice of the system.

Alternatively, one can say that $LM +$ is a *quasi-Boolean algebra* (Rasiowa 1974), for this is by definition an algebra which is a distributive lattice, and in which a unary operation ' has the properties of being an involution and satisfying the de Morgan identities (i.e. ' meets conditions C2 and C3). Lemmas 3.11, 3.14 and 3.15 show that this quasi-Boolean algebra is also a distributive lattice in which every element has a relative orthocomplement. I do not think this property of the representative algebra has been recognised before, and so the classical nature of this three-valued logic has been unrecognised.

Stated in the form above, theorem 3.2 makes clear in exactly what sense LM is a classical logic. As noted earlier, the inferential structure of LM is essentially the same as in CBL, since each logic is represented by a distributive lattice. That is to say that the connectives $\supset$, $\vee$, $\cdot$ and $\equiv$ in either system are all essentially the same. From theorem 3.2 we see that negation too is fundamentally the same, since in either logic it is represented by an orthocomplement. The only difference is that, while in the bivalent logic this is a full lattice orthocomplement, in the non-bivalent system it is relative to a sublattice of the system. The scope of negation in LM is restricted, so that it has its full classical properties only within a particular subset of wffs which we call the relevant formulae. In the bivalent logic there is no such restriction. In CBL it is assumed that the context of each wff is the entire logic, or alternatively that relevance is a trivial relation.

3.2.2 *The representation of* LM$_T$

The representation theorem 3.2 applies to the logic of any theory. For any theiry T the logic LM$_T$ is represented by an algebra $LM_T +$ which is at least a quasi-Boolean algebra. $LM_T +$ may of course have extra properties since the laws of a theory may impose extra conditions on the valuations of a theory, and so there may be tautologies of LM$_T$ which are not valid in LM. However, all tautologies of LM are also valid in each LM$_T$, and so the representative algebra $LM_T +$ has at least the properties of $LM +$ discussed above. The construction of $LM_T +$ for any T is now briefly presented.

In order to construct algebra $LM +$, recall that we can distinguish the tautologies of this logic from those of LM, writing $T \models \alpha$ when α is true in every valuation of T. One can define a relation of theoretical equivalence in T in the obvious way:

$$\alpha R \beta \qquad \text{iff} \quad T \models \alpha \equiv \beta.$$

All wffs which are logically equivalent in LM are also theoretically equivalent in this sense. One can show that R is an equivalence relation in the required sense, by an argument analogous to that used for CBL and LM. Thus one can generate the equivalence classes

$$[\alpha] = \{\beta : T \models \alpha \equiv \beta\}.$$

The carrier set of LM$_T +$ is therefore

$$\text{LM}_T + = \{[\alpha] : \alpha \text{ is a wff of LM}_T\}.$$

The relations $\leqslant$ and $'$ on LM$_T +$ which represent logical connectives in this logic LM$_T$ are defined by

$$[\alpha] \leqslant [\beta] \qquad \text{iff} \quad T \models \alpha \supset \beta$$
$$[\alpha]' = [\neg \alpha].$$

One shows these relations are well-defined by arguments similar to those given for CBL and LM above. In a similar way one can also show that $\leqslant$ is indeed a partial ordering, and one can introduce lattice operations with respect to $\leqslant$ which represent disjunction and conjunction in LM$_T$:

$$[\alpha] \bigvee [\beta] = [\alpha \vee \beta]$$
$$[\alpha] \bigwedge [\beta] = [\alpha \cdot \beta].$$

By arguments analogous to those above, one can show that $LM_T + =$

$\langle \text{LM}_T +, \leqslant, ' \rangle = \langle \text{LM}_T +, \vee, \wedge, ' \rangle$ is a quasi-Boolean algebra. As above, each wff has a context, a subset of relevant wffs. Algebra LM_T is at least a distributive lattice in which every element has a relative orthocomplement.

3.2.3 Summing up

The algebra representing logic LM was constructed using the method discussed earlier in chapter 2. We saw that this is a distributive lattice, in which every element has a relative orthocomplement. This indicates that implication, disjunction, conjunction and the biconditional are essentially the same in either logic. Negation too is essentially the same, in the sense that it has the properties of an orthocomplement. However, in CBL this is a full lattice orthocomplement, while in the non-bivalent logic LM it is relative to a subsystem of relevant wffs.

Since LM has only the properties common to the logic of any mechanical theory, these remarks apply equally to any individual logic LM_T. Such a logic is essentially classical.

4 States and Measurement in Mechanics

4.1 The states of a theory

4.1.1 *What is a state?*

Recall from §1.1 that a state in classical mechanics corresponds to a point in phase space, and hence to an assignment of values to every magnitude of the theory. It follows that a classical state also corresponds to a two-valued valuation of the theory, for a two-valued valuation finds every simple proposition either true or false, and so an atomic proposition of the form (m, r) is found to be true for every magnitude m, thus assigning a particular value r to each magnitude. In classical theories, states and two-valued valuations coincide.

In general, however, one cannot define a state as a two-valued valuation, for some theories may lack them altogether. We saw that the elementary set E_T of any T is a poset which is at least a union of Boolean algebras. It may not be possible to map such a set consistently onto the two-element set $\{t, f\}$ (see §3.1). In fact, Kochen and Specker (1967) have argued that theories of quantum mechanics do not generally have bivalent valuations. At any rate, a definition of state which is appropriate to all theories cannot assume the existence of these functions. Two-valued valuations should arise as a special case of a general definition.

Intuitively, we want states to be valuations which express the most information possible given by a theory. Those states which make the most truth-value assignments are said to be the 'fullest' valuations. In this sense they express the 'most complete' descriptions of reality. These conditions are made precise if we require that a state is a valuation which has a maximal 'truth-set' of simple propositions. In order to give a formal definition, one must introduce some new terms.

First, consider the set H_T of valuations of a theory T. Associated with each valuation h in H_T are the following sets of wffs in the logic LM_T.

Definition
The *elementary truth-set* of h, $ETh = \{p : \models_h p\}$.
The *truth-set* of h, $Th = \{\alpha : \models_h \alpha\}$.

The elementary truth-set of h contains all the simple propositions true in this valuation, while the truth-set contains any wff in the logic which is true in h. Analogous sets of false propositions and wffs can also be defined:

Definition
The *elementary falsity-set* of h, $\text{EF}h = \{p: \dashv_h p\}$.
The *falsity-set* of h, $\text{F}h = \{\alpha: \dashv_h \alpha\}$.

It follows from the definition of $\neg$ that $\text{EF}h$ contains the negations of all the propositions in $\text{ET}h$, and vice versa. Similarly, $\text{F}h$ contains the negations of all wffs in $\text{T}h$. Given one of these sets, the others can be determined.

In fact, one can show that any valuation is uniquely characterised by either truth-set:

Lemma 4.1
$h_1 = h_2$, iff $\text{ET}h_1 = \text{ET}h_2$, iff $\text{T}h_1 = \text{T}h_2$.

First, note that $\text{ET}h_1 = \text{ET}h_2$ iff $\text{T}h_1 = \text{T}h_2$. This follows from the fact that the rules defining connectives in the logic LM_T uniquely extend any simple valuation of propositions, to a unique complex valuation of wffs (see §2.1). Lemma 4.1 is established by noting that when two truth-sets coincide, so do the two falsity-sets, and hence also the sets of undecided wffs. This is just to say that the two valuations must coincide on all assignments of values, and so the two valuations are the same. It follows from this lemma that valuations can be represented by their elementary truth-sets or their truth-sets or even corresponding falsity-sets in a completely unambiguous way.

Relations between elementary truth-sets can therefore generate relations between corresponding valuations. In particular, one can define

Definition
h_2 contains h_1, $h_1 \leqslant h_2$, iff $\text{ET}h_1 \subseteq \text{ET}h_2$.

One valuation in H_T contains another if it makes at least the same assignments of truth to simple propositions. One shows that $\leqslant$ is indeed a partial ordering on H_T, the set of valuations of theory T, by appeal to the properties of $\subseteq$. For example, since $\text{ET}h \subseteq \text{ET}h$, $h \leqslant h$, and so $\leqslant$ is reflexive.

A state of T can now be defined in terms of the containment relation. A state will be maximal with respect to this partial ordering:

Definition
h is a *state* of theory T if h is a valuation of T and if there is no other h^* in H_T, $h^* \neq h$, such that $h \leqslant h^*$.

No other valuation can contain a state. No other valuation can make the same truth-value assignments to simple propositions, and yet make more.

In this sense the states of a theory express its most detailed, most 'complete' descriptions. One immediate consequence of this definition is that, where they exist, bivalent valuations will always be states:

Lemma 4.2
If h is two-valued, h is a state.

This follows simply from the definitions above. Clearly, no other valuation can contain a bivalent assignment of truth-values, and so these will always be states. It is most important to note, however, that the converse of this lemma does not generally hold. Nothing in the definition of a state requires that it be a bivalent valuation.

However, where they do exist, bivalent states turn out to be important, so we give a special name to the following theories:

Definition
T is *classical* if all states of T are bivalent.

States of classical theories will be called *classical states*. Theories and states without these properties will be called *non-classical*. Since nothing in the definitions requires bivalence, states and theories are generally non-classical. In later discussion of probabilities we shall see that these terms have their usual meaning.

4.1.2   *Representing states and valuations*

We now determine how the states and valuations of any theory can be represented algebraically. In fact, lemma 4.1 will be used in the algebraic analysis below. First, some additional concepts from modern algebra must be introduced.

The fundamental concept is that of a filter:

Definition
A *filter* F of lattice A is a proper subset of A which has the following properties:

> 1   If $a \in F$, and $b \in F$, then $a \wedge b \in F$.
> 2   If $a \in F$, and $a \leqslant c$, then $c \in F$.

Thus a filter is a non-empty subset of A which is not the whole set, which is closed with respect to meet and which contains every element of the lattice which is 'above' any member of the filter (see Bell and Slomsen (1974) p 12, Sikorski (1969) p 12 and Rasiowa (1974) p 42). Some important properties which a filter F may have are given in the following standard definitions:

Definition
F is *prime* if, when $a \vee b$ is in F, either $a \in F$, or $b \in F$.

F is *maximal* (an *ultrafilter* of A) if there is no other filter F^* of A such that $F \subset F^*$.

In any lattice with an involution $'$, one can also define:

Definition
F is *contradictory* if, for some a, $a \in F$ and $a' \in F$.
F is *non-contradictory* if, whenever $a \in F$, $a' \notin F$.
F is *exhaustive* if, for any a, either $a \in F$ or else $a' \in F$.

The relevance of these terms to logic will now become apparent.

First, we construct the subset of algebra $L M_T +$ which represents the truth-set of a valuation h of T, and hence the valuation itself.

Definition
$$Th+ = \{[\alpha] : \models_h \alpha\}.$$

This is the set of equivalence classes representing wffs in Th, i.e. representing the wffs which are true in h. The set of all such sets for a theory T is

$$H_T+ = \{Th+ : h \in H_T\}.$$

Clearly, this represents the set H_T of valuations of theory T.

We now return to the algebraic representation of a valuation:

Lemma 4.3
$Th+$ is a filter of LM_T+.

$Th+$ is clearly a subset of LM_T+. To show that it is a proper subset, note first that it must be non-empty, since very h at least makes truth-assignments to logically true wffs of LM_T. It is not the whole of LM_T+, since the negations of true wffs will not be true in h, and so their equivalence classes will not be contained in $Th+$. $Th+$ is closed with respect to meet by definition of conjunction and lemma 3.9. Finally, $Th+$ contains every equivalence class 'above' any member, since whenever $[\alpha] \in Th+$ and $[\alpha] \leqslant [\beta]$, then $\alpha \in Th$ and $\models \alpha \supset \beta$, and so $\beta \in Th$ also by the definition of h, and so $[\beta] \in Th+$.

In fact, not only is $Th+$ a filter of LM_T+, it also has some special properties.

Theorem 4.1
For any valuation h in H_T, $Th+$ is a prime and non-contradictory filter of LM_T+.

By lemma 4.3, $Th+$ is a filter of LM_T+. The properties of being prime and non-contradictory follow from the definitions of the logical connectives $\vee$ and $\neg$ in LM_T. For example, if $[\alpha] \vee [\beta] \in Th+$, then $\models_h(\alpha \vee \beta)$, so by the definition of $\vee$, either $\models_h \alpha$ or $\models_h \beta$, and so $[\alpha] \in T_h+$, or $[\beta] \in Th+$ and so the filter is prime. A similar argument shows that $Th+$ is non-contradictory, and so the theorem is proved. Note that the converse of this theorem does *not* hold. Not every prime non-contradictory filter of LM_T+ will represent a valuation of theory T, for such a filter may correspond to an assignment of truth-values which is inconsistent with the laws of theory T.

One can now determine how the states of any theory T will be represented. In fact we conclude no more than has already been presented. The states of T will be represented by prime non-contradictory filters $Th+$ in lattice LM_T+, where h is maximal with respect to the containment relation. According to this analysis it is not the case, as is often assumed, that states are represented by maximal filters in LM_T+, for maximal filters of LM_T+ will generally be contradictory, and so cannot represent the truth-set of any valuation, let alone a state. A filter F of LM_T+ may, for example, include an entire *m*-subset of simple propositions, so that negations p, $\neg$p will be in F for any *m*-proposition p. Such a filter F may be a proper subset of LM_T+, and may include the filter $Th+$ which represents a state.

In fact, states are not even characterised by filters which are maximal in the subset H_T+ of LM_T+, for a filter $Th+$ may be maximal in this subset H_T+, and yet fail to represent a state. This follows essentially from the fact that logic LM_T has operators expressing the relation of any wff to any valuation. Suppose for example that h_1 is a valuation which is not a state, so that $h_1 \leqslant h_2$ for some other h_2 in H_T. This is to say that for some proposition p in E_T, $h_1(p) = u$, but $h_2(p) = t$. In this case, wff Up is true in h_1, but Up is false in h_2, so that although $ETh_1 \subseteq ETh_2$, $Th_1 \not\subseteq Th_2$, and so Th_1+ may indeed be maximal in the set H_T+, even though h_1 is not a state[†].

Traditionally, it has been supposed that states are represented by maximal filters. In fact, this view stems from some very special results which characterise Boolean algebras. The traditional analysis of classical states, which is founded on these results, will now be briefly examined.

4.1.3 States and the logic of a theory

According to the traditional view, classical theories use a classical bivalent logic, and their states are represented by maximal filters in the Boolean algebra representing this logic. One can agree that an analogous result does hold in the special case of classical theories. Classical theories do not actually use a classical bivalent logic to express their descriptions, but one can associate with any classical theory T a bivalent 'logic of states' LSM_T which has this structure. This logic is generated from the bivalent valuations of the theory. It can be shown that the states of T will

---

† Recall that U is the logical connective defined in §3.1.4, so that Uα is true only when α is undecided, and is false otherwise.

be represented by maximal filters of the algebra LSM$_T$+ representing this bivalent logic. That is:

Lemma 4.4
If T is a classical theory, and h a state in H$_T$, then Th+ is an ultrafilter of the Boolean algebra LSM$_T$+ .

That LSM$_T$+ is a Boolean algebra follows from earlier discussion of CBL+ , in §2.2. =he correspondence between bivalent valuations and Boolean ultrafilters is a well-known result, which follows from some special properties of filters in Boolean algebras (see e.g. Bell and Slomsen (1974) or Sikorsky (1969)). In particular, it follows from the fact that for filters of a Boolean algebra the properties of being prime, of being both non-contradictory and exhaustive and of being maximal, all coincide. This special Boolean result probably explains the fact that states and maximal filters are often assumed to correspond.

Traditionally, a special Boolean theorem is considered to be at the heart of any discussion of classical states. This is Stone's Representation Theorem for Boolean algebras:

Stone's Theorem
Any Boolean Algebra A is isomorphic to a perfectly reduced field of sets, by a mapping which associates with each elemtn a of the algebra set X(a) of ultrafilters of A which contain it.

The field of sets X(a) is known as the *Stone Space* of algebra A†.

This theorem is traditionally counted as important, for at least two reasons. First, since classical states are supposed to be represented by Boolean ultrafilters, Stone's Theorem is supposed to characterise the classical space of states. Each equivalence class in LSM$_T$+ is associated, by Stone's isomorphism, with the set of ultrafilters which contain it. Since these ultrafilters are taken to represent the truth-sets of states of T, each wff in the logic LSM$_T$ is supposed to be associated, by Stone's Theorem, with the set of states in which it is true.

Secondly, the theorem is traditionally supposed to characterise the logical structure of a classical theory. Recall that traditionally it is assumed that a bivalent 'logic of states' expresses the descriptions of a classical theory. Since this is a classical bivalent logic, it is represented by a Boolean algebra, and so this theorem is supposed to reveal the topological structure of the logic of a theory. It tells us that the algebra representing the bivalent logic LSM$_T$ has the structure of a perfectly reduced field of sets.

These two reasons for finding Stone's Theorem of interest yield, in fact, a third. It is usually considered significant that the very same theorem

† The topological definitions do not concern us here, but full details of the theorem can be found, for example, in Bell and Slomsen (1974) or Sikorski (1969).

characterises the structure both of the classical space of states and of the logic used by a classical theory. This is supposed to reveal a fundamental relation between the states and logic of classical theories, so that the structure of states can be assumed to be just the structure of a theory's logic. As one modern author writes (Bub 1974 p 106):

> ... the phase space of a classical mechanical system is a topological characterisation of the propositional structure: the phase space is the Stone Space of the Boolean algebra of propositions.

It has seemed very puzzling that no such relation between states and logic is evident in quantum theories. It is well known that quantum states cannot be represented in terms of a perfectly reduced field of sets, and there are various reasons for supposing that the logic of quantum theories cannot be a strictly Boolean logic. Attempts have been made to represent quantum states by maximal filters of some other kind of algebra, which might represent both the space of quantum states and a non-classical 'quantum logic'. However, this approach has had no great success. The traditional analysis of classical states and logic does not seem to apply to quantum theories.

However, the classical results need very careful re-appraisal. The structure of states in a theory is essentially different from the structure of its logic, and only the very special properties of Boolean filters have led us to suppose that these are the same. The essential properties of filters which represent truth-sets are that they are prime and non-contradictory, but in general, prime non-contradictory filters are not maximal. It is only in Boolean systems that the two characterisations are equivalent. In any theory, a state is maximal with respect to the containment relation, but, as discussed earlier, this does not mean it is represented by a maximal filter. States and ultrafilters do not correspond.

As already explained, in the special case of a classical theory one *can* construct a special 'logic of states' LSM_T, and hence a Boolean algebra $\text{LSM}_\text{T}+$, which does have ultrafilters representing states. However, even in classical theories, this 'logic of states' LSM_T is not the logic expressing all descriptions of the theory, for descriptions are expressed even when there are truth-value gaps (see the earlier discussion in §3.1.3). I would argue that, even in this classical case, the structure of states and the algebraic structure of a theory's logic are not the same.

As far as a topological representation of the logic of a theory is concerned, there is a theorem for LM which is analogous to Stone's theorem for CBL. This is the representation theorem for quasi-Boolean algebras, and is due to Bialynicki-Birula. This theorem shows that any quasi-Boolean algebra is isomorphic to a 'quasi-field' of open subspaces of a certain compact space†. It is interesting that the isomorphism is

---

† As with the Stone Space, the topological details are not important here, but full details of this theorem can be found in Rasiowa (1974) p 44.

established by a mapping which associates, with each element a of the algebra, the set $X(a)$ of prime filters which contain it. By analogy with Stone's theorem one might call this space of sets $X(a)$ the B-B Space of the algebra. Since prime and maximal filters coincide in Boolean algebras, this is indeed the analogue of the Stone Space defined above.

It is this theorem rather than Stone's which topologically characterises the logic used by any theory, since it is LM rather then CBL which is the logic of mechanical theories. Clearly, however, this theorem does not characterise the structure of states in a theory, for, as we saw earlier, prime filters of an algebra LM_T+ may be contradictory, and so do not generally represent states. The structure of states of a theory will not be the structure of its logic. Once the distinction is made between these two structures we can properly understand classical theories and above all we can understand the similarities between classical and quantum mechanics.

4.1.4 Summing up

In the first section we saw that the states of any theory T are special valuations. They are maximal with respect to containment, making the most truth-value assignments to elementary propositions. Next it was shown that valuations are represented by prime and non-contradictory filters in the algebra representing their logic. In general, states are not characterised algebraically as maximal filters; indeed, they are not even maximal in the subset of filters which do represent valuations.

Since, traditionally, maximal filters and states have been supposed to correspond, the traditional analysis of classical states and classical logic was briefly discussed in §4.1.3. This analysis assumes that a classical bivalent logic is used by classical theories. Because of special properties of Boolean filters, the structure of classical states and the logic of a theory appear to be the same. However, in general, a non-bivalent logic is used, and we see that these structures are distinct even in the case of classical theories.

The notion of a 'phase space' of states will be re-examined in the discussion of probabilities, in chapter 6. In the meantime we turn to the role of measurement in mechanical theories.

4.2 The role of measurement

4.2.1 The logical role of measurement

Here we consider the logical properties of measurement—the role which it plays in description. We shall ignore its physical properties. Obviously, measurement does involve a physical interaction with reality, and clearly in some mechanical theories the nature of this interaction may be interesting, but I hope to show that the peculiar features of quantum theories, which are often attributed to physical disturbance, can in fact be explained purely in terms of the logical properties of measurement. The physical properties need not be considered.

Thus here measurement will be characterised logically, in a way that applies to any theory no matter what its subject matter. In fact, it seems fundamental to the logical role of measurement that it establishes the truth-value of some propositions. After all, we perform measurements to obtain information about the values of magnitudes, and this information is expressed when truth-values are assigned to simple propositions. We therefore define:

Definition

A *measurement* M in theory T is some procedure, in principle specifiable, which results in the assignment of truth to some proposition in E_T.

The procedural nature of measurement is not important, although we assume that it can be specified. The essential logical property is that a measurement M performed on the reality described by T establishes the truth of some proposition of this theory.

It should be noted here that truth and falsity are primitive terms in this discussion, which will not be defined. The nature of description, and the philosophical question of how propositions can truly describe a reality, are not investigated. In the next section, certain minimal assumptions will be made about description, but the fundamental metaphysical issues are not addressed. The logical analysis may inform our metaphysics, but we do not wish the converse to be the case. Metaphysical assumptions will not underlie the logical analysis.

The definition of measurement allows us to introduce several useful terms. First is the notion of a measurement outcome:

Definition

An *outcome* of measurement M is any simple proposition which is true after M has been performed.

Trivial cases are included in the definition since, for example, there is no requirement that an outcome is not also true before the measurement has taken place.

We shall let h_M be the valuation which describes a reality after a measurement M is performed. From the definition of M above, and the notion of an outcome, we conclude that the truth-set of h_M contains all the outcomes of M, and all their valid consequences in the theory. One way of expressing this is to say that h_M contains the valuation which is characteristic for these outcomes. To express this precisely one defines for any wff α of LM:

Definition
h_α is the *characteristic valuation for* α if

$$Th_\alpha = \{\beta \in \text{LM} : T \models \alpha \supset \beta\}.$$

h_α is thus the valuation which finds α true and all valid consequences of α in the theory true, but which makes no further truth-assignments. One can now conclude that

Lemma 4.5
If p is an outcome of M, then $h_p \leqslant h_M$.

This follows simply from the definitions, and could be taken as another statement of what it means for a proposition to be a measurement outcome.

The definition of a measurement outcome also allows us to distinguish several different kinds of measurement. For example:

Definition
M is a *measurement* of P if it is a measurement which has either p or $\neg$p as an outcome.
M is an *m-measurement* (a measurement of magnitude m) if it has some m-proposition as an outcome.
M is an *atomic measurement* if it has some atomic proposition (i.e. of form (m, r), for r a real number) as an outcome.

We can also say that M is a *measurement of* α if it is a measurement of every constituent proposition in α. Where M is a measurement of m, we may sometimes use the symbol m to refer to the measurement as well as the magnitude.

We see from this discussion that a measurement is essentially an individual procedure, which is characterised by its outcomes. Thus it is also essentially theory-dependent, for its outcomes are propositions, and these clearly depend on the theory considered, in particular on its magnitudes and value-sets. It is quite possible that exactly the same physical process would be regarded as a different measurement in theories which described the same physical system using different magnitudes. For example, if we use the term 'mass' in theory T, then certain procedures will be regarded as measurements of mass in this theory, since they are understood as providing truth-assignments to mass propositions. However, if some other theory describing the same reality had no such magnitude, there could clearly be no measurement of mass in this theory. Perhaps the same procedure would be understood differently as a measurement of some quite different magnitude.

4.2.2 *The successor relation*

Before a measurement takes place, we say that a system is described by an *initial* valuation h in H_T. A subsequent measurement M of the system may provide new truth-values which were not assigned initially. Thus after measurement this initial valuation may be replaced by a different valuation in H_T which describes the same physical system. We understand the logical role of measurement when we understand this relation, between any initial valuation and others which may follow it after measurement. This binary relation will be called the *successor* relation. It relates any valuation h to those other valuations in H_T which might be used immediately after h to describe the same reality.

At least two assumptions can be made about the successor relation, and since these are minimal assumptions, common to our understanding of measurement in any mechanical theory, they will be taken as conditions which define it. First we assume that no successor of h can actually contradict its elementary truth-value assignments. If h finds a simple proposition p to be true, then a successor of h, used immediately after h to describe the same system after measurement, will not actually find that p is false. This minimal logical condition, which we assume to hold in any theory at all, seems to follow from our acceptance of h as an initial description of reality.

Of course, one might argue that measurements do alter reality, and so a contradictory valuation might indeed be used to describe a system after measurement. However, as remarked above, we are not concerned here with any physical properties of measurement. Instead, we seek to determine the logical properties common to measurement in any theory, irrespective of its physical properties. We therefore assume that measurements are non-disturbing, just as we shall assume that they can be performed arbitrarily quickly, so that no evolution of the system takes place. Actual cases, where measurements may disturb and time elapse, should be analysed using additional principles, which take these extra properties into account. However, we hope to show that the logical properties of measurement will by themselves explain many interesting properties of quantum theories.

Using the terms introduced earlier, one can express the assumption that succeeding valuations are not contradictory, in a precise way:

Non-contradiction of successors
If h_1 is an initial valuation, and h_2 is used immediately after h_1 to describe the same reality after a measurement has taken place, then $ETh_1 \cap EFh_2 = \varnothing$.

There can be no simple proposition which is true in h_1 but false in h_2, if these valuations are to be successive descriptions of the same reality, divided only by a measurement.

The second assumption about successive valuations is that, where possible, information is conserved throughout the process of measurement. We do not simply 'lose' preceding truth-value assignments of simple propositions after a measurement has taken place. According to this assumption, where new truth-values can be consistently added to those of earlier valuations, they are so added. This can be expressed as the following formal condition, using the earlier definition of a characteristic valuation h_α:

Conservation of information
If h_1 is an initial valuation, and h_2 is used immediately after h_1 to describe the same reality after a measurement, then for every p in ETh_1, if there is an h^* in H_T, such that $ETh^* = ETh_2 \cup ETh_p$, then $h_p \leqslant h_2$.

If p and all its consequences do not conflict with the truths of h_2 they will also be true in this successive valuation. Where possible, initial truth-assignments to simple propositions are retained.

These two conditions are combined to define, formally, the successor relation on H_T.

Definition
h_2 is a *successor* of h_1 in H_T, $h_2 S h_1$, iff
   1   $ETh_1 \cap EFh_2 = \varnothing$, and
   2   For every p in ETh_1, if there is an h^* in H_T, such that $ETh^* = ETh_2 \cup ETh_p$, then $h_p \leqslant h_2$.

This relation is now used to express, in a formal way, the important logical role of measurement. We know already that h_M finds all outcomes of M to be true. In addition one now concludes the following:

Principle of succession
After a measurement M is performed on a physical system initially described by h, the same reality will be described by a successor of h, h_M (i.e. $h_M S h$).

After M, a system is described by a successor h_M of h. We know from the earlier discussion of measurement that h_M will not generally be identical to h, since a measurement has outcomes which may not have been true initially. This principle, and lemma 4.5 above, give the logical properties of measurement.

It is worth remarking that much stronger conditions than 1 and 2 above could have been used to define the successor relation. However, 1 and 2 do seem minimal, uncontroversial conditions which we can safely assume of any theory. Furthermore, we shall show that, although these conditions may seem very weak, they are sufficient in classical theories to generate familiar, stronger properties.

4.2.3 Classical and non-classical succession

In classical theories, where states are always bivalent, it turns out that there is a very special relationship between initial valuations and their successors. In these theories, the successor relation coincides with containment:

Lemma 4.6
$h_2 \, \mathrm{S} \, h_1$, iff $h_1 \leqslant h_2$, in any classical theory.

Recall that $h_1 \leqslant h_2$ means that $\mathrm{ET}h_1 \subseteq \mathrm{ET}h_2$, so according to this lemma one valuation succeeds another in a classical theory only if it shares the elementary truth-value assignments of the first. This result follows from the fact that all states are bivalent, for whenever there are valuations h_1 and h_2 which are such that $\mathrm{ET}h_1 \cap \mathrm{EF}h_2 = \varnothing$, there must be an h^* which is such that $\mathrm{ET}h^* = \mathrm{ET}h_2 \cup \mathrm{ET}h_\mathrm{p}$ and so whenever h_2 is a successor of h_1, $h_1 \subseteq h_2$. All simple truth-values can be retained in successive valuations, in a classical theory.

Thus classical theories turn out to be a very special case, where the valuations used to describe a system after measurement always agree with initial simple truth-value assignments. In this special classical case we can always understand measurements as 'discovering' new truth-values, which are then simply added to previous truth-value assignments. Information simply accumulates until a complete 'picture' of reality is established. Thus a series of valuations $h, h_1, \ldots, h_n$ in a classical theory, describing a reality initially and after a succession of measurements M_1, $\ldots, M_n$, is always a series whose elementary truth-sets increase, so that $\mathrm{ET}h \subseteq \mathrm{ET}h_1 \subseteq \cdots \subseteq \mathrm{ET}h_n$. Such a series will ultimately end with a classical state, since it follows from lemma 4.6 above that bivalent valuations will have only themselves as successors. We therefore understand the sequence of measurements in a classical theory as providing successive determinations of some unique classical state. After the state is reached, we know that measurement will provide no further new truth-values, and we therefore conclude that our information, our description of the reality, is 'complete'.

Thus lemma 4.6 attributes a special significance to the classical state. Since only the classical states have no non-trivial successors, they are assumed to be in unique correspondence with reality. The only other valuations describing the same reality as a state are the non-bivalent valuations contained by it. Non-bivalent valuations in classical theories are understood as 'incomplete' descriptions of a reality which is fully characterised by some unique classical state.

This unique relation between a bivalent state and the reality it describes has led to widespread ambiguity in the use of the term 'state'. The same word is traditionally used to refer both to the maximal state-descriptions of a theory, and to the physical states-of-affairs which these describe. In classical theories, the ambiguity is not too confusing, just because we assume the unique association of states with realities. However, in non-classical theories we shall see that this association is not unique, and so it becomes vital to make the distinction. Paradoxes arise in non-classical descriptions if we fail to distinguish properly the state-descriptions of a theory from physical states-of-affairs.

In non-classical theories, lemma 4.6 above does not hold. Succession in these theories does not coincide with containment, and so our simple classical understanding of states and measurement must be carefully revised. Successive measurements in these theories cannot be understood as successive 'discoveries' of new truth-values which then simply accumulate until a final 'picture' is reached. In a series $h, h_1, \ldots, h_n$ of valuations in a non-classical theory, describing a reality initially and after successive measurements $M_1, \ldots, M_n$, the relation of containment will not generally hold. Although there may be some cases in the series where $h_i \leqslant h_{i+1}$, and so $\mathrm{ET}h_i \subseteq \mathrm{ET}h_{i+1}$ for $1 \leqslant i < n$, in general this is not the case. It may be impossible to combine successive elementary truth-values into a single consistent elementary truth-set. That is, it may be impossible to find an h^* such that $\mathrm{ET}h^* = \mathrm{ET}h_k \cup \mathrm{ET}h_{k+1}$, $1 \leqslant k < n$, and so $\mathrm{ET}h_k \not\subseteq \mathrm{ET}h_{k+1}$. In such a case, since h_{k+1} S h_k, these valuations do not actually assign contradictory truth-values to elementary propositions. But since $h_k \not\leqslant h_{k+1}$, there will be at least one simple proposition p which is true in h_k but undecided in h_{k+1}. The information expressed by p is in this sense 'lost' because it cannot be consistently combined with the other truth-values of h_{k+1}. It may not be possible to combine the truth-values in a way that is consistent with the logic and laws of a non-classical theory.

Thus in general, where states are not always bivalent, a sequence of valuations describing a reality initially and after a series of measurements will not be a sequence whose elementary truth-sets simply increase. In general, we can conclude only that each valuation is a successor of its immediate predecessor. This condition ensures that such a sequence of descriptions must always be *locally* consistent, in the sense that no valuation contradicts an elementary truth-value of its immediate predecessor. However, the sequence may not be *globally* consistent since one valuation may contradict other members further removed from it in the sequence. For instance, in a sequence of descriptions, a simple proposition p of a non-classical theory may be true in h_k, undecided in h_{k+1}, and may actually be false in some later valuation h_i, $i > k+1$.

We have seen above, from lemma 4.6, that the special properties of classical theories ensure that these are always globally consistent. All preceding elementary truth-values are preserved by later valuations in a sequence of successors. Non-classical theories, however, do not have this special property. Where states are not bivalent, information cannot always be consistently combined, and so may be 'lost', and perhaps eventually contradicted in this way. This is a result of the weak structure of non-classical theories, and in particular of their states. This structure does not generate a strong successor relation.

We see here too the reduced significance of the non-classical state. In a non-classical theory, where lemma 4.6 does not hold, one cannot assume a unique correspondence between states and the physical realities they describe. A non-classical state may have non-trivial successors, so that more than one state might be used after measurement to describe the same undisturbed reality. It therefore becomes important to distinguish the states of a theory, which are just its maximal valuations, from the physical states-of-affairs described. One is no longer the unique 'picture' of the other.

This is not to say, however, that a non-classical theory leaves us in straightforward 'ignorance' of the state-of-affairs—a state does not merely omit some part of a complete description. For instance, we cannot say, when state h finds all atomic m-propositions undecided, that 'in state h the physical system described by the theory really has a value for m, but we do not know what it is'. Instead we must question the very meaning of the terms 'a physical system really has an m-value' and avoid this kind of statement, especially where non-classical theories are concerned. It is the magnitudes of a theory which 'really have values', and this is the same as saying that atomic propositions 'really have truth-values'. This much we assume in the descriptive relation between a theory and reality. However, we should not assume more about the nature of reality itself. After all, we can speak with authority of our theories and their logic of description, but not of the nature of reality. Where theories are non-classical we must be especially careful in our distinctions, or paradox will result. We can say of a state that 'magnitude m takes value r', or that 'proposition (m, r) is true', but we should avoid the metaphysical claim that 'the physical system has m-value r'. In a classical theory this can be taken as synonymous with the earlier statements, but in a non-classical theory it cannot, because of the lack of bivalent states.

Thus, in non-classical theories, our choice of magnitudes does not provide a simple 'picture' of the system described. States assigning truth-values to propositions about one magnitude may be unable to decide propositions about other magnitudes. These states leave us in ignorance of reality in the sense that we do not have a bivalent description, and in this sense lack a 'picture' of the system. However, this is not an 'ensemble' interpretation, i.e. an 'ignorance' interpretation in the usual sense. We cannot assume that some 'larger' theory might have 'larger' states assigning values to all the magnitudes of a non-classical theory T.

In this discussion of succession, no account has been taken of the physical nature of measurement. One assumes that measurements are performed ideally quickly, so that no evolution takes place, and also with ideally little disturbance. Hence the same unaltered system is assumed to be described before and after measurement. The change in state which may occur in non-classical theories after measurement is not describing a real change in the physical state-of-affairs. It is simply made necessary by the impossibility, in non-classical theories, of combining all elementary truth-values in a consistent way. A change in state after measurement indicates that the theory concerned is non-classical, and so when new information about a system is introduced, old information may be lost. In classical theories information can always be consistently combined, and such changes in state will not occur. This point will be discussed again in chapter 6, where probabilities are considered.

4.2.4 Compatibility

There are of course cases, even in non-classical theories, where measurement will not lead to a change in the state describing a reality. In order to discuss such cases, one can introduce the following logical definitions:

Definition
Wffs α, β of LM_T are *compatible relative to* h (compatible in h) iff they are both decided by h (i.e. iff $h(\alpha) = t$ or $h(\alpha) = f$ and also $h(\beta) = t$ or $h(\beta) = f$). If one wff is decided but the other is undecided in h, they are *incompatible relative* to this valuation.

α, β are *absolutely compatible* if there is no state in which they are incompatible. The wffs are *absolutely incompatible* if they are not compatible relative to any state of T.

Thus two wffs are compatible in h when both of them, separately, are either true or false in this valuation. The wffs are incompatible when one is either true or false, but the other is undecided. They are absolutely compatible when they are always decided in the same states. Two wffs are absolutely incompatible when no state can decide them both. For example, two m-propositions p and q will be compatible in any state h, and so are absolutely compatible.

These definitions are extended to the compatibility of magnitudes:

Definition
Magnitudes m, n are *compatible relative to* h if every pair of m- and n-propositions is compatible relative to h. These magnitudes are *incompatible relative to* h if some pair of m- and n-propositions is incompatible relative to this valuation.

m, n are *absolutely compatible* if there is no state in which they are incompatible. These magnitudes are *absolutely incompatible* if they are not compatible relative to any state of the theory.

Two magnitudes are compatible relative to h when all their m- and n-

propositions are true or false in this valuation, and are incompatible in *h* if some are true or false, but others undecided in *h*. The magnitudes are absolutely compatible if all their simple propositions are always decided in the same states, and are absolutely incompatible if no state of the theory can decide all of these propositions.

It follows immediately from the definitions that in a classical theory all wffs and all magnitudes are absolutely compatible, for all states of a classical theory are bivalent and so decide every wff in the logic of the theory. However, in theories which are not classical, this is not generally the case. Where states are not always bivalent, wffs and magnitudes may not be absolutely compatible. This feature of non-classical theories is evident in quantum mechanics. We shall now show that compatibility, as defined above, has the characteristic properties of compatibility as it occurs in quantum theories.

Although it is not always recognised, there is in quantum mechanics a sense of relative as well as of absolute compatibility (see e.g. Hardegree 1977). Recall that in quantum theories compatibility can be defined by two equivalent conditions (see §1.1). Two observables *A* and *B* of quantum theory T are absolutely compatible if the operators which represent them in the Hilbert Space associated with T share an eigenbasis, or alternatively if these operators commute. However, one can also say that *A* and *B* are compatible relative to the state ψ if this state vector is common to the eigenbasis of $\hat{A}$ and that of $\hat{B}$, or alternatively if operators $\hat{A}$ and $\hat{B}$ commute with respect to this vector—i.e. if $\hat{A}\hat{B}\psi = \hat{B}\hat{A}\psi$. Clearly, these conditions may be met even though *A* and *B* are not absolutely compatible.

Compatible observables in quantum theories are often said to be 'simultaneously measurable'. This may refer to the fact that both can be assigned a precise value with probability 1, in a single state of the theory. This property of compatible observables will be discussed later, where probability assignments are analysed (see chapter 6). However, one can also say that the observables are 'simultaneously measurable' in the sense that the order in which measurements of the observables are performed on a system will not alter its final state. It is this property which we now consider.

First one can show:

Lemma 4.7
If *m* and *n* are compatible relative to valuation *h*, then the order of measuring these magnitudes on a system initially described by *h* will not affect the final state-description.

This follows simply from the definitions. Since *m* and *n* are compatible in *h*, any pair of *m*- and *n*-propositions is decided in this valuation. Thus the outcomes of both measurements will already be true in *h*, and so this valuation will also describe the system after both measurements, irrespective of their order. In addition to this, we have the more interesting result:

Lemma 4.8
If m and n are absolutely compatible, then the order of measuring these magnitudes will never affect the final state-description, whatever the initial valuation.

Again this follows from the definitions. Since m and n are absolutely compatible, so are their outcomes. It follows that there is a state of the theory which finds the outcomes of both measurements true, and so no information will be lost during the measurement process. To put this more exactly, suppose h is the initial valuation, M a measurement of m and N a measurement of n. If M is performed first, then we know that the valuation describing the system afterwards is h_M, where $h_M \, \mathrm{S} \, h$, and h_M finds the outcomes of M to be true (by lemma 4.5 and the Principle of Succession). If N is performed immediately afterward, the system is then described by h_{MN}, where $h_{MN} \, \mathrm{S} \, h_M$ and h_{MN} finds the outcomes of N true. But since the outcomes of the two measurements are compatible, and cannot be contradictory by the definition of succession, we also know that the final valuation h_{MN} retains the truth-values assigned to outcomes of M, by condition 2 defining succession. Hence this valuation has, in its elementary truth-set, all outcomes of M and all of N, together with any simple propositions found true by h which do not conflict with these truth-assignments. A similar argument establishes that the same properties characterise valuation h_{NM}, which results from measuring the magnitudes in the opposite order. Thus these valuations are the same.

In non-classical theories there will of course be magnitudes which are not compatible, and so the final state of a system will generally vary with the order in which measurements are performed. However, in classical theories, where magnitudes are always compatible, the order of measurement will never alter the state describing a system. Once more we stress that this is a difference in the logical properties of measurement in the theories, for physical properties have not been considered.

This difference in the logical role of measurement is not a difference in propositional logic of the theories. We have seen that classical and non-classical theories of mechanics use essentially the same propositional logic—in either case it has the structure of LM. The fundamental difference between these theories is in the properties of their states. It is the bivalence of classical states which gives classical theories, and in particular classical measurements, their special properties. To understand this as a difference in the logic of the theories, one must consider their *modal* logic. This will be discussed in chapter 7 below. In the meantime, we turn to the question of classical and non-classical probability assignments, for probability statements too can be expressed in the modal logic of a theory.

4.2.5 Summing up

Measurements were characterised as physical procedures, in principle specifiable, which establish the truth of simple propositions. These propositions are called the outcomes of the measurement. It was remarked that the procedural nature of measurement, i.e. its physical properties, is not considered here. It is the logical properties which concern us, for these are sufficient to explain many peculiar features of quantum mechanics.

The logical properties of a measurement M can be investigated if we examine the relation between the initial valuation h, describing the reality before M is performed, and the resulting valuation h_M, describing the system after M. This is called the *successor* relation, and it was defined by two fundamental assumptions. First, we assume non-contradiction—i.e. that successive valuations do not actually contradict preceding elementary truth-assignments. Secondly, we assume that information is conserved—i.e. that earlier truth-values of simple propositions are retained where this is possible.

In classical theories these two fundamental properties generate much stronger conditions. In fact, where all states are bivalent, succession coincides with containment, so a successor of h agrees with all truth-value assignments made by h to simple propositions. A sequence of classical valuations describing a system after measurements is therefore a sequence whose truth-values simply increase until some final state is reached, for classical states have only themselves as successors, and so correspond uniquely to the reality described. For this reason, the single term 'state' can be used classically, without great ambiguity, to refer both to the maximal valuations of a theory, and to the physical state-of-affairs which these describe. We tend to assume that the state is a 'picture' of the state-of-affairs.

In non-classical theories, however, succession is a much weaker relation. These theories do not have the special properties which make succession coincide with the containment relation. Hence a sequence of measurements does not generate a sequence of valuations whose elementary truth-assignments merely accumulate until some bivalent 'picture' is established. Instead, earlier information may be 'lost', and even contradicted later in the sequence. In this sense, non-classical theories are locally, but not globally, consistent.

Compatibility was defined by a logical condition, and we showed that, where magnitudes are compatible, the state of a system will not vary with measurement, but where they are incompatible the state of a system may be changed after measurement. To understand this situation we must be careful to distinguish states, i.e. the maximal valuations of a theory, from physical states-of-affairs. Non-classical states do not correspond uniquely to realities, and so states may be changed without an alteration in the state-of-affairs. Changes in state after measurement stem from the logical properties of measurements which are assumed to be non-disturbing. These changes indicate that the non-classical successor relation is weak, and that information may be lost as well as gained by measurement.

Finally, the different logical properties of measurement in classical and non-classical theories do not indicate a difference in their propositional logic, for all mechanical theories use essentially the same logic LM. Instead, this difference in the role of measurement stems from the different properties of states in either theory. It must, therefore, be expressed in a logic which includes a modal operator. Such a logic is discussed in chapter 7 below.

5 The Traditional Analysis of Probabilities

5.1 Introduction to Kolmogorov probabilities

5.1.1 Defining probabilities

A set X of elementary 'events' is fundamental to the traditional analysis of probabilities. Probabilities are numbers which have certain special properties. They are assigned to subsets of X, in order to indicate how 'likely' the corresponding events are to occur. We are not concerned here with the philosophical foundations of probability theory, and so the notion of 'likelihood' remains undefined. Instead the formal foundations of probabilities are to be analysed, in particular the fundamental relation between logic, states and the probabilities assigned by a theory. First we examine the general definition of probabilities.

To take a simple example, consider a throw of a die[†], and let set X represent the possible outcomes of this experiment. Since any one of its six faces may be upturned after the die is thrown, we consider X to have six distinct members which we label $e_1, \ldots, e_6$ according to the number $1, \ldots, 6$ which appears on the upturned face. Since we use probability 1 to indicate a certainty, the probability or 'weight' of each member of X, $P(\{e_i\})$ will generally be less than 1. For instance, we say that the die is unbiased if all sides are equally likely to appear, in which case for each i, $1 \leqslant i < 6$, $P(\{e_i\}) = \frac{1}{6}$. If the die were biased we might have a much greater probability associated with some particular outcome: for example, it could be the case that $P(\{e_2\}) = \frac{1}{2}$, indicating that the face with number 2 is much more likely that the others to be upturned after a throw.

Set operations on X are supposed to generate corresponding relations among the probabilities. Using the example above of an unbiased die, we can consider the probability of some 'compound event', for example the appearance of an even number after the die is thrown. Here we are considering the event represented by subset $\{e_2, e_4, e_6\} = \{e_2\} \cup \{e_4\} \cup \{e_6\}$, and the probability is calculated by adding the associated probabilities, so that $P(\{e_2, e_4, e_6\}) = P(\{e_2\} \cup \{e_4\} \cup \{e_6\}) = P(\{e_2\})$

[†] See, for example, Gemignani (1970) for an exposition of some simple examples.

$+ P(\{e_4\}) + P(\{e_6\}) = \frac{1}{2}$. Where, as in this case, the subsets involved have no members in common, set union corresponds to the addition of probabilities.

A similar relation exists between the operation of set complement, and the subtraction of probabilities. Again, using the earlier example, suppose that we wish to calculate the probability that neither of the numbers 2 or 3 appears on the upturned face of the die. This involves finding the probability of the compound event $\{e_1, e_4, e_5, e_6\}$. By the above discussion, we know that this could be calculated as $P(\{e_1\}) + P(\{e_4\}) + P(\{e_5\}) + P(\{e_6\})$, using the correspondence between set union and addition, but alternatively we could find it by noting the relation between set complement and subtraction: $P(\{e_1, e_4, e_5, e_6\}) = P(X - \{e_2, e_3\}) = 1 - \frac{1}{3} = \frac{2}{3}$.

The theory of probabilities was first presented as a rigorous mathematical system in the 1930s, by Kolmogorov. In the simple case where set X is assumed to be a finite set, Kolmogorov gave the following axioms for the theory of probabilities (Kolmogorov 1956):

Let E be a collection of elements ξ, η, ζ which we shall call elementary events, and F a set of subsets of E; the elements of set F will be called random events.

1   F is a field of sets.
2   F contains the set E.
3   To each set A in F, is assigned a non-negative real number $P(A)$. This number $P(A)$ is called the probability of the event A.
4   $P(E) = 1$.
5   If A and B have no element in common, then $P(A + B) = P(A) + P(B)$.

A system of sets, F, together with a definite assignment of numbers $P(A)$ satisfying axioms 1–5, is called a *field of probability*.

Kolmogorov defines a field of sets as a system which is closed with respect to union, intersection and complement. An additional 'Axiom of Continuity', imposing an additional condition on the field F, is given for the case where set X may be infinite.

Since Kolmogorov's pioneering work, the theory of probabilities has become just one part of the more general mathematical theory of measures. Following the methods of measure theory, Kolmogorov's axioms are generally separated into those concerning the structure of *F*, containing subsets of some fundamental set X, and those concerning the function which assigns numbers to these subsets.

Definition
F, a set of subsets of X, is a *σ-algebra*, if

1   $X \in F$
2   If $A \in F$, then the set complement of A, $-A \in F$
3   For any sequence $\{A_n\}$ *of F*, the set union $\cup_{n=1}^{\infty} A_n \in F$.

Thus F is a σ-algebra if it contains X and is closed with respect to set complement and countable unions. A probability measure over F is now defined in two stages[†].

Definition
μ is a *measure* over F, if μ is a real-valued function from F which is such that:

1   $\mu(0) = 0$
2   $\mu(A) \geqslant 0$, for all A in F
3   For any sequence $\{A_n\}$ of F, whose members are pairwise disjoint (i.e. $A_i \cap A_j =$, for all $i \neq j$, $1 \leqslant i, j \leqslant n$), $\mu(\cup_{n=1}^{\infty} A_n) = \sum_{n=1}^{\infty} \mu(A_n)$

P is a *probability measure* over F, if P is a measure over F and

4   $P(X) = 1$.

Thus a measure is a function which assigns real numbers to sets in F, assigning 0 to the empty set and non-negative real numbers to all other members of F, and which is σ-additive (condition 3). A probability measure has, in addition, the property that set X is assigned 1. All other members of F will be assigned numbers in the interval $[0, 1]$ by a probability measure.

Using these terms, one defines

Definition
A (Kolmogorov) *probability space* is a triple (X, F, P) where X is some set, F a σ-algebra of subsets of X and P is a probability measure over F.

When X is a finite set, we say the probability space is *finite*, and when X is infinite, the space is *infinite*. Here, finite spaces will generally be used for discussion and illustration. This is not because they are the most common—indeed, mechanical theories often use infinite probability spaces, but the finite spaces make discussion simpler. Generalisation to infinite probability measures is usually straightforward, involving the accepted methods of measure theory. Where a probability space (X, F, P) is finite, the probability measures can be expressed as weighted sums, as in the earlier example of a die. For any A in F one can write $P(A) = \sum_{x_i \in A} w_i$. Here the weights w_i are just the probabilities assigned to members of X, $P(x_i) = w_i$.

---

† For the modern measure-theoretic approach to probabilities see, for example, Bauer (1972) or Renyi (1970).

5.1.2 Some properties of probabilities

The familiar properties of probability assignments can easily be deduced from the definitions, or from Kolmogorov's axioms†. For example, the correspondence between set complement and subtraction, which was mentioned in an earlier example, can be derived from these principles.

Lemma 5.1
$P(-A) = 1 - P(A)$ for A any set in F.

Sets A and $-A$ are, by the definition of $-$, disjoint sets, and so the result follows from axioms 4 and 5 of Kolmogorov, or from conditions 3 and 4 in the definition of a probability measure.

One can also derive the familiar properties of 'conditional' probabilities, from the principles above. Traditionally, one defines

Definition

$$P(A/B) = P(A \cap B)/P(B), \quad \text{if } P(B) \neq 0$$
$$= 0 \quad \text{otherwise.}$$

Traditionally $P(A/B)$ is called the conditional probability of A given B, though later we shall refer to it as a *ratio* probability (see §6.1.2). This can be understood as the probability of A, given that we are restricting our attention to outcomes which also have the property B. One can easily show that this function is itself a probability measure, both over the whole σ-algebra F generated from X, and also on that subalgebra F_B of F which is generated from the subset B of X. We let P_B be the probability defined above, $P_B = P(A/B)$. Then

Lemma 5.2
If $B \subseteq X$, and F_B is the σ-algebra generated by B, then

1. (X, F, P_B) is a probability space
2. (B, F_B, P_B) is a probability space.

Both results are established by checking that the appropriate axioms hold. Another familiar result concerning this kind of probability is the following:

Lemma 5.3
If $A \cap B = 0$, then $P(A/B) = 0$.

If the events have no element in common, their conditional, or ratio probability is zero. Indeed, all the familiar results of probability theory can be derived from the axioms of Kolmogorov, or from the relevant principles of measure theory.

---

† The proof of this and subsequent lemmas can be found in any standard text, for example, Gemignani (1970), Bauer (1972) or Renyi (1970).

In addition to the simple spaces (X, F, P) associated with a set X, one can also generate *joint* probability spaces, to assign probabilities to several different events. These can be generated either from one single space, as when we consider several similar experiments, or else they can be constructed from different fundamental sets.

Consider, for example, the earlier case of a throw of a die, with set X = $\{e_1, \ldots, e_6\}$ representing the six possible outcomes. Suppose we now throw the die twice, and wish to calculate the probabilities of outcomes from this experiment. The elements of interest now are ordered pairs of elements in the original set X. For example (e_1, e_3) represents the 'event' of finding 1 on the upturned face after the first throw, and 3 after the second. The new set generating a probability space for these events is X^2, the cartesian product of X containing all ordered pairs of its elements. The new σ-algebra F^* of subsets of this set is generated by set operations on X, for example $(e_1, e_2) \cup (e_3, e_4) = (e_1 \cup e_3, e_2 \cup e_4)$. Finally, the probability measure P^* on the probability space is defined by the condition that $P*(A, B) = P(A) \times P(B)$, where P is the original probability measure. This probability measure P^* calculates the joint probability of independent events A and B. Using this construction of a joint probability space, one can for example calculate the probability that number 2 appears on the first throw of a die, and 5 on the second, with the argument that $P*(\{(e_2, e_5)\}) = P(\{e_2\}) \times P(\{e_5\}) = \frac{1}{6} \times \frac{1}{6} = \frac{1}{36}$.

It is easy to check that a joint space (X', F', P') generated from the cartesian product of a set X is indeed a Kolmogorov probability space. In a similar way one can construct probability spaces which assign joint probabilities to events from different sets. Suppose for example, that the sets X_1, X_2 are associated with probability spaces (X_1, F_1, P_1), (X_2, F_2, P_2), respectively. One can generate from these a joint, or product, space (X^*, F^*, P^*) which assigns probabilities to events (x_1, x_2), where $x_1 \in X_1$ and $x_2 \in X_2$. Here X^* is the product $X_1 \times X_2$, F^* is the σ-algebra generated from this product by set operations on each set, and P^* is a probability measure over this algebra. Where subsets A_1, A_2 of F_1 and F_2 represent independent events, then $P^*(A_1, A_2) = P_1(A_1) \times P_2(A_2)$.

Traditionally, much of the discussion of probabilities in mechanics is expressed in terms of random variables. Briefly, we understand a random variable as expressing some property, some aspect, of the events which might occur. One might be interested, for example, not in a particular number upturned after the throw of the die, but in whether or not an even number appears. This property can be represented by a function f from the set X of possible outcomes, to the set $\{1, 0\}$. f takes each e_i representing an even number to 1, and all other elements of X to 0. Alternatively, we might be interested in the number of odd numbers appearing in two throws, in which case we can define a function f^* from the product set X^* discussed above to the set $\{0, 1, 2\}$, which takes each ordered pair to the number of odd numbers represented. For example, $f^*(e_1, e_4) = 1$, $f^*(e_2, e_6) = 0, f^*(e_3, e_5) = 2$, and so on.

Formally, one defines a random variable on any probability space as follows:

Definition

f is a *random variable* on the probability space (X, F, P), if *f* is a real-valued function $f: X \rightarrow \mathbb{R}$, which is measurable with respect to F (i.e. $f^{-1}(B) \in F$, for every Borel set B in $\mathbb{R}$).

The requirement that *f* be measurable with respect to *F* arises because we wish to assign probabilities to the ranges of values of a random variable. That is, one may wish to calculate the probability that *f* takes a value in subset B of its range, and for such a probability to be well-defined, we require that the events mapped to this interval are in the σ-algebra *F*.

The function which does assign probabilities to values of *f* is called a probability distribution, and is defined as follows:

Definition

The *probability distribution* of random variable *f* on X is the function P^f on the family of Borel sets of $\mathbb{R}$, defined by the condition that $P^f(B) = P(f^{-1}(B))$, for each Borel set B.

Corresponding to this, one can define a *probability density* function, which assigns probabilities to individual values in the range of *f*. One can check that P^f is indeed a probability measure over the σ-algebra of Borel subsets of $\mathbb{R}$. According to the definition, the probability that *f* takes a value in B, $P^f(B)$, is just the probability of the event $f^{-1}(B)$, $P(f^{-1}(B))$.

We saw that joint probability spaces can be constructed to define joint probabilities of events, either from the same or from different probability spaces. An analogous construction can be applied to random variables. For example, if *f* and *g* are two random variables over spaces (X_1, F_1, P_1), (X_2, F_2, P_2), then the joint probability distribution of f, g, P^{fg}, is a probability measure over pairs of Borel subsets of $\mathbb{R}$, which is derived as follows. A joint probability space (X, F, P), where $X = X_1 \times X_2$, is defined according to the earlier discussion. The composition $f \cdot g$ of the random variables *f* and *g* is defined over the cross product X in the usual way so that $f \cdot g(A, B) = (f(A), g(B))$, for A, B, Borel subsets of X_1, X_2 respectively. The inverse of this map is $(f \cdot g)^{-1}$, where $(f \cdot g)^{-1}(A, B) = f^{-1}(A) \cdot g^{-1}(B)$. The joint probability distribution of these two random variables is then obtained by setting $P^{fg}(A, B) = P((f \cdot g)^{-1})(A, B))$. This construction can be generalised to give joint probability distributions for *n* random variables.

We can understand $P^{fg}(A, B)$ as the probability which tells us how likely it is that the value of the random variable *f* is in A, and the value of *g* is in B. As remarked, in the case of probabilities, joint distributions are traditionally used also to calculate sequential probabilities, as we shall later discuss.

Finally, one often wishes to determine the average, or 'expected', value which a random variable may take. Formally this is defined:

Definition
The *expected value* of random variable f is $\langle f \rangle = \int_x f(x) P(x) \, dx$, where this integral exists.

Where set X is finite, the integral is of course a weighted sum of values of the random variable. That is, if $X = \{x_1, \ldots, x_n\}$, and $P(x_i) = w_i$, $1 \leqslant i \leqslant n$, $f(x_i) = r_i$, then $\langle f \rangle = \sum_i w_i r_i$. Derived from this definition is the expression for the RMS (root mean square) deviation Δf of a random variable f.

Definition
$$\Delta f = (\langle (f^2 - (f)^2) \rangle)^{1/2}$$

Δf indicates the dispersion of f, its 'scatter' about the expected value.

5.2 The traditional analysis of classical mechanics

Here we shall see how the methods above are traditionally applied to classical mechanics. In the next section we discuss the problems raised when we try to apply these methods to quantum mechanics. In the next chapter a new approach is proposed which retains the traditional definitions of probability theory, but applies them to mechanics in a different way.

5.2.1 The event space

Recall that any classical theory of mechanics can be represented by means of an underlying 'phase space', the space being parametrised by its generalised position and momentum variables (see §1.1.1 above). We saw, for example, that a classical theory describing a system with n degrees of freedom will use a $2n$-dimensional phase space of 'points' $(q_1, \ldots, q_n, p_1, \ldots, p_n)$, where the q_i are position variables and the p_i are momentum variables respectively. According to traditional analysis, these points represent the fundamental 'events' of the theory. The set X of such points generates the fundamental probability space (X, F, P) used by a classical theory. Here F is the σ-algebra generated from X by means of set operations, and P is a probability measure over F. (X, F, P) is understood as the '*event space*' of the theory, the probability space over which all its probabilities are defined.

We saw earlier that the points in classical phase space represent the states of the theory (§1.1.1 and §4.1). Furthermore, when the precise state of a system is unknown, it can be described by a 'probability state', a probability measure over the space (X, F, P). The probability assigned to each point in X indicates how 'likely' the system is to be in this state.

This traditional analysis of classical theories can be expressed in terms of random variables. Again, the probability space (X, F, P) generated from the phase space is taken as fundamental. Observables of a classical theory are represented by random variables on this probability space. For example, in a theory which has a two-dimensional phase space, position will be represented by the function f_Q which takes any point (q_i, p_j) to its position value q_i: $f_Q(q_i, p_j) = q_i$. Similarly, all position and momentum observables can be represented by random variables on X. Since all other magnitudes of the theory are functions of these, they too will each correspond to a random variable.

Thus each magnitude of a classical theory is associated with a probability distribution. For any magnitude m, represented by random variable f_m on X, one can calculate the probability $P^m(B)$, that the value of m lies in the set B (recall that $P^f(B) = P(f^{-1}(B))$). Furthermore, one can associate an expected value $\langle (f_m) \rangle$ with every magnitude, according to our earlier definition, and one can calculate the RMS deviation Δf_m, the 'uncertainty' of m, which expresses the dispersion of the values of m about its expected value.

Joint probability distributions can be defined for any magnitudes $m_1, \ldots, m_n$ in a classical theory. The joint probability distribution $P^{m_1, \ldots, m_n}$ is constructed according to the earlier definitions so that, for example, $P^{m_1, \ldots, m_n} (B_1, \ldots, B_n)$ gives the probability that the value of magnitude m_1 is in B_1 and that the value of m_2 is in B_2, and so on. In classical mechanics, joint distributions are also used to calculate the probabilities in a sequence of measurements. For example, the expression above might be used to determine the probability that a measurement of m_1 yields an outcome in B_1 and then a subsequent measurement of m_2 yields value B_2, and so on, where we assume the measurements are 'ideally' close in time, in the sense that no evolution of the system occurs between them. In the next chapter this use of joint distributions to calculate sequential probabilities will be examined in detail.

5.2.2 Probabilities and logic

It is often assumed that this representation of classical theories, using the set X of points in phase space and its associated 'event space' (X, F, P), is easily related to discussion of logic. In fact, the relation between (X, F, P) and logic has seemed such a trivial matter that generally no explicit discussion is given. Authors pass happily from talk of probabilities of 'events' to discussions of probabilities of propositions or of classical states, without ever bothering to make the fundamental relations among these concepts clear. However, our primary interest here is to understand these relations, in particular to understand how the classical analysis is generalised in quantum theories. It therefore becomes crucial that we examine the assumptions relating events, propositions and logic properly in the traditional analysis of classical mechanics.

The first traditional assumption, which indeed seems well justified, is that a unique relation exists between points in the phase space and atomic propositions in the logic of a classical theory. This seems justified just because the points $(q_1, \ldots, q_n, p_1, \ldots, p_n)$ do assign particular real values to the observables of a theory. For example, in a theory using a two-dimensional phase space, a point (q_i, p_j) assigns the value q_i to the magnitude position and the value p_j to momentum. An atomic proposition (m, r) is also fundamentally an assignment of some particular value r to a magnitude m of the theory. Thus we can say that the point (q_i, p_j) in the set X of the theory T represents the ordered pair $((Q, q_i), (P, p_j))$ of atomic propositions from E_T, where Q is the position magnitude and P the momentum magnitude in this theory. In this way, the set X in the event space of a theory which uses a two-dimensional phase space corresponds to an analogous set X_p of ordered pairs of atomic propositions. To simplify the discussion, one can simply regard set X as the set of all values of observables in the theory, and hence the corresponding set X_p will be the set of all atomic propositions in E_T.

Since the set X_p of atomic propositions exactly corresponds to the set X of points in phase space, the probability space (X_p, F_p, P_p) generated by X_p exactly corresponds to the event space (X, F, P) which was generated from X. The space of propositions and the space of events are equivalent. Here F_p is simply the σ-algebra of subsets of propositions in X_p generated by set operations. Clearly, F_p has exactly the same structure as F, the algebra of sets in the event space of the theory.

The next traditional assumption is that the σ-algebra F_p of sets of propositions represents, in fact, the logic of the classical theory. It is therefore assumed that set operations represent logical connectives, but recall from §2.2 above that it is classical bivalent logic, CBL, which is represented by a Boolean algebra of sets. By assuming that F_p represents the logic of a theory, one assumes that a classical theory uses a classical bivalent logic.

The last important step in the traditional analysis is to associate with the spaces (X, F, P) of events and (X_p, F_p, P_p) of propositions a third equivalent space (X_s, F_s, P_s) of Boolean ultrafilters which are supposed to represent the states of the theory. This probability space is derived from the others by means of Stone's Theorem, already discussed in §4.1.3 above. Recall that this theorem establishes an isomorphism between any Boolean algebra and its Stone Space, a field of Boolean ultrafilters, in such a way that every element of the algebra is mapped to the set of Boolean ultrafilters which contain it. The theorem can be applied either to the algebra F in the event space of a classical theory, or to the algebra F_p in the probability space of propositions. Either way, the resulting Stone Space of ultrafilters F_s will be the field in a probability space (X_s, F_s, P_s). This can be understood as a probability space of states, wince as we have seen already (§4.1.3), these ultrafilters can be used to represent states of a classical theory. The measure P_s assigning probabilities to sets of ultrafilters in the Stone Space F_s can be understood as a measure of the states in which an event of F occurs, or alternatively a measure of the states in which a proposition in F_p is true. The probability assignment P_s corresponds to the measures P and P_p which assign probabilities directly to events or propositions.

Thus traditionally, at least, three probability spaces turn out to be equivalent. The event space (X, F, P) generated by the phase space X of the theory, the corresponding space (X_p, F_p, P_p) generated from the set X_p of atomic propositions, and lastly by means of Stone's isomorphism, the space (X_s, F_s, P_s) of Boolean ultrafilters which represent the classical states. Traditionally, it is assumed that the space of events (X, F, P) or of propositions (X_p, F_p, P_p) is fundamental, while the space of states is derived from these. In the next section, we shall see some problems encountered by this view. In the next chapter a different approach is suggested, where one takes instead the probability space of states as fundamental to a theory.

5.3 The problem of quantum probabilities

5.3.1 The problem

The traditional analysis of classical theories, outlined above, seems inapplicable to the probabilities of quantum mechanics, for quantum theories appear to have no analogue of the classical phase space X of points, and so have no analogue of the classical event space (X, F, P), which was taken as fundamental to classical theories. Indeed, it appears at first that there is no probability space at all which generates the probability assignments of quantum theories.

The following example illustrates the problem. This example, discussed by Bub (1974), concerns a quantum theory T which is associated with a three-dimensional Hilbert Space, $\mathcal{H}_T$. A, B, are two incompatible magnitudes of the theory, represented by operators which have discrete, non-degenerate eigenvalues, a_1, a_2, a_3 and b_1, b_2, b_3 respectively, and corresponding eigenvectors $\alpha_1, \alpha_2, \alpha_3$, and $\beta_1, \beta_2, \beta_3$. According to quantum mechanics, the following probabilities are associated with state α_1:

$$prob_{\alpha_1}(a_1 : A) = |(\alpha_1, \alpha_1)|^2 = 1$$
$$prob_{\alpha_1}(a_2 : A) = |(\alpha_2, \alpha_1)|^2 = 0$$
$$prob_{\alpha_1}(a_3 : A) = |(\alpha_3, \alpha_1)|^2 = 0.$$

The probability according to state α_1 that a measurement of magnitude A will yield outcome a_1 is 1, and the probability assigned by this state to all other outcomes of the same observable is 0. The same state α_1 also makes non-zero probability assignments to the outcomes of magnitude B. That is,

$$prob_{\alpha_1}(b_1 : B) = |(\beta_1, \alpha_1)|^2 \neq 0$$
$$prob_{\alpha_1}(b_2 : B) \neq |(\beta_2, \alpha_1)|^2 = 0$$
$$prob_{\alpha_1}(b_3 : B) \neq |(\beta_3, \alpha_1)|^2 = 0.$$

These probabilities raise a problem for the traditional analysis. For instance, suppose we calculate the joint probability that A has the value a_1, and B the value b_1. By hypothesis, magnitudes A and B are incompatible, and so there is no state of the theory in which both are assigned a precise value. It follows that there is no state of the theory in which the joint event is obtained, or in which the corresponding conjunction of propositions $(A, a_1) \cdot (B, b_1)$, is true, and so according to the traditional definition this joint probability must always be zero. Thus from this it follows that the quantum probabilities above cannot represent classical conditional probabilities, for, according to the traditional definition, a conditional probability is always zero when the corresponding joint probability is zero (see lemma 5.3). So here we have the following problem: the joint probability of (A, a_1) and (B, b_1) is zero, and yet quantum mechanics tells us that the conditional probability of (B, b_1) given (A, a_1), i.e. $prob_{\alpha_1}(B : b_1)$ that B has value b_1 given state α_1 is non-zero. How should one understand such quantum conditionals?

A second problem concerns the sequential probabilities of quantum mechanics. Recall that, traditionally, the probability assigned to a sequence of measurement outcomes is calculated by finding their joint probability (see §5.2.1). However, in quantum mechanics this construction does not in fact give the probability of a sequence, for reasons similar to those above. One can, for example, have non-zero probabilities for sequences of events whose joint probabilities are zero.

Consider the earlier example, where A and B are incompatible magnitudes of a quantum theory. Suppose a sequence of three measurements M_A, M_B, M_A^* is performed on a system described by this theory, in such quick succession that the system does not evolve in the process. Here A is measured first, then B, and lastly A is measured again. It is a well-known feature of quantum mechanics that in such a case one does not expect the same outcome from the two measurements of A. For instance, a_1 might be the outcome of the first measurement M_A, but this does not mean we predict a_1 with certainty as the outcome of the third measurement M_A^*. This is a simple consequence of von Neumann's Projection Postulate, discussed in §1.1.4. After M_A, which has outcome a_1, the Projection Postulate tells us that the system is left in state α_1, the eigenvector corresponding to eigenvalue a_1. The second measurement, M_B of B, will have an outcome b_i, and hence will leave the system in state β_i, according to the same principle. Thus in calculating the probability that the third measurement M_A^* will find value a_1 of A, we must calculate the probability $prob_{\beta_i}(a_1 : A) = |(\alpha_1, \beta_i)|^2 \neq 1$. The outcome a_1 is not certain according to quantum theory.

Now consider the case where the measurements are performed in a different order. For example, consider the sequence $M_B M_A M_A^*$, where observable B is measured first, followed by two successive measurements of A. Again we assume that no evolution of the system takes place. In this case it is possible to predict the outcome of the second measurement of A with certainty, for according to the Projection Postulate, the first measurement leaves the system in state β_i, and after the second measurement this changes to state α_1, corresponding to the value a_1 of A. Hence, now, when we wish to calculate the probability that the same value is found after a second measurement of A, we calculate $prob_{\alpha_1}(a_1 : A)$ $= |(\alpha_1, \alpha_1)|^2 = 1$. According to quantum mechanics we are now certain to find the same outcome a for the measurements M_A and M_A^*.

This example shows that sequential probabilities are order-dependent in quantum theories, for we see here that $prob_h \langle a_1 : A, b_1 : B, a_1 : A \rangle$ (i.e. that value a_1 is found after a measurement of A, then b_1 after a subsequent measurement of B, and lastly a_1 again after another measurement of A) is not the same as $prob_h \langle b_1 : B, a_1 : A, a_1 : A \rangle$, where the measurements are performed in a different order. Similarly, the probability assigned to value a_1 of A, given that this magnitude is measured, varies according to the order of measurements we suppose to have taken place beforehand. However, this order-dependence, so clearly evident in the probabilities of quantum theories, is certainly not a property of joint probability assignments or of probabilities derived from these. It appears that quantum sequential probabilities cannot be derived from the traditional definitions which are used in classical theories. These definitions are used successfully where all magnitudes are compatible, but where incompatible magnitudes are involved they are inadequate.

5.3.2 *Some reactions to the problem*

It is impossible here to review the vast amount of work which has been done, and is still being done, in the foundations of quantum mechanics and probability theory in attempts to resolve these problems. The subject is interesting but involves complex issues and would require a volume of its own. (A detailed historical survey up till the early 70s can be found for example in Jammer 1974.) At most I can merely outline some differences in approach to the problem, and can indicate how my own analysis, presented in the next chapter, differs from these.

One reaction to these problems has been the search for 'hidden variables', which can be added to existing quantum mechanics in order to 'complete' the theory. Essentially this is an attempt to find new magnitudes, value-sets and laws so that new propositions and relations are generated. In this way it is hoped that an augmented probability space can be constructed which will generate quantum probabilities according to traditional methods. Many attempts at constructing this kind of hidden variable theory have been made, and are discussed for example in Jammer (1974) or Bub (1974). There are similarly many different 'proofs' which purport to show that such attempts are impossible. The work of Kochen and Specker (1967) has already been mentioned as demonstrating that the elementary system E_T of a quantum theory T cannot be consistently assigned to a two-element algebra. This seems to indicate that even an augmented set E_T will not in fact admit bivalent valuations if the structure of the present theory is preserved. At any rate, there seems to have been no persuasive success for this approach, in the sense that no analogue of classical phase space, with the traditional classical properties, has been found to generate the probabilities of quantum theory.

One reaction to the problem, which is currently popular, is to conclude that some aspect of the mathematical foundations of probability theory fails to apply in the quantum case. Most authors lately have argued that at least one of Kolmogorov's axioms, or correspondingly one of the modern principles of probability theory, fails in the foundation of quantum probabilities. A great deal of detailed work has been done in this area which I cannot hope to review[†]. It is usually concluded that the σ-algebra F has a weaker structure with corresponding alteration in the properties of probability measure P defined over it. So far, this approach too has not had great success, and clearly it introduces problems of its own. For example, in what sense are quantum assignments truly probabilities if the traditional conditions do not hold?

In some cases this change in the axioms of quantum probabilities stems from an argument about quantum events, or properties. Thus, for example, Bub (1974) writes

> If the property theory is non-classical—i.e. non-Boolean or non-set-theoretical—then the probability theory on these properties has to

[†] A brief review of some of this work can be found in Chari (1971).

be non-classical too. We require a generalisation of the classical Kolmogorov theory of probability appropriate to a strongly non-Boolean possibility structure.

Alternatively this can be seen as an argument about the logic of quantum theories. The lack of a Kolmogorov probability space is supposed to show that the logic of these theories is fundamentally non-classical, for if there were a Boolean algebra representing the logic of the theory, this could take the role of the σ-field in a fundamental probability space. Thus Suppes, for example, writing in a well-known article in 1966, argues that the lack of a joint probability space is (Suppes 1966)

> ... the single most powerful argument for the use of a non-classical logic in quantum mechanics.

In the following chapters a different approach to quantum probabilities is proposed. The usual axioms for probabilities are retained but are applied in a different way to theories of mechanics. In particular, there is no fundamental event space which generates all probabilities of a quantum theory. The set X of events is not taken as fundamental, nor is a corresponding set of propositions. Instead, a set of states generates the probability spaces of a theory. The probability of a wff is to be a measure of the set of states of the theory in which this wff is true. We saw that in classical theories this analysis could be derived from the traditional notion of an event space; here it is taken as fundamental. In the next chapter we shall see that this provides a formal foundation for probabilities in mechanics, one in which probabilities are traditional Kolmogorov measures, and yet will be shown to have the characteristics of quantum assignments in non-classical theories.

6 The Probabilities of Mechanics

6.1 The new foundation for probabilities in mechanics

6.1.1 Predictive conditional probabilities

Traditionally, one talks about probabilities of events, of wffs, or of states, without drawing any careful distinction between them. The probability space of events is usually taken as fundamental, and spaces of propositions and states are derived from this. Here, however, this traditional analysis is abandoned. We are concerned with the logical foundations of probabilities, and so with the probabilities of wffs, or of states. The term 'event' is ambiguous, except perhaps in the special classical case, and will not generally be used. The probability of a wff is to be a measure of the set of states in which this wff is true. Thus the fundamental probability spaces of a theory are spaces generated from a set of states. We shall see, however, that other spaces, of sets of wffs, propositions or values of magnitudes can be derived from these.

In fact, a conditional probability function is taken as fundamental. This is the predictive conditional, $prob_h(\alpha : M)$, expressing the 'likelihood' that wff α is true after measurement M is performed on a system initially described by h. Once more the term 'likelihood' is undefined, since we are investigating the formal foundations of probabilities and leaving the philosophical questions to one side (cf comments in §5.1.1). The probability function above is called *predictive*, since it gives the likelihood of α being true after a measurement M. We are therefore considering the valuations which might describe the system after h. It is a *conditional* probability in the sense that it depends on the initial valuation h, as well as on M.

To give a formal definition of the probability measure, recall the discussion in chapter 4 of states and measurement. In particular, recall that after a measurement M is performed on a system, the initial valuation h may be changed to another valuation h_M, which is a successor of h, and which finds the outcomes of M to be true (see §§4.2.1 and 4.2.2). In discussing probabilities, however, we are generally interested in cases where the particular outcome of a measurement procedure M is unknown. In such cases, we can associate a set of possible outcomes with the procedure M, and hence a set of valuations which might result from the measurement:

Definition
β_M is the *characteristic wff* for M, if it is the disjunction of all wffs which might be outcomes of M.

For example, if M is a measurement of m, β_M will be the disjunction of all m-propositions. Wff β_M is characteristic, in the sense that after M a system will be described by a valuation h_M, in which this wff is true.
   Using this method of characterising measurement procedures, we can identify the set of states which will generate the probability space of a predictive conditional probability. For any initial valuation h of a theory T, and any measurement M, one can define:

Definition
The set of M-*successor* states of h is $St_{hM} = \{h' : h'$ is a state of T, and $\models_{h'}\beta_M\}$.

This is the set of states which might be used to describe a system initially described by h after measurement M has been performed. The set generates a probability space as follows. First, one defines the subset of states which find a particular wff α true, and then one constructs the field of all such subsets:

Definition
$$St_{hM}^{\alpha} = \{h' \in St_{hM} : \models_{h'}\alpha\}$$
$$F_{hM} = \{St_{hM}^{\alpha} : \alpha \in LM_T\}.$$

St_{hM}^{α} is the subset of states in St_{hM} which find wff α true. F_{hM} is the set of all such subsets, for all wffs in the logic of a theory.

Lemma 6.1
$F_{hM} = \langle F_{hM}, \cup, \cap, - \rangle$ is a field of subsets of St_{hM}.

   Clearly, F_{hM} is a set of subsets of St_{hM}. By definition of the logic, wff $(p \supset p)$ for example, is true in every valuation of T, and so $St_{hM} = St_{hM}^{p \supset p}$ is in F_{hM}. To show that F_{hM} is closed with respect to set operations, we show that these correspond to appropriate logical connectives. That is, one shows that $St_{hM}^{\alpha} \cup St_{hM}^{\beta} = St_{hM}^{\alpha \vee \beta}$ and $St_{hM}^{\alpha} \cap St_{hM}^{\beta} = St_{hM}^{\alpha \cdot \beta}$, and also $- St_{hM}^{\alpha} = St_{hM}^{\sim\alpha}$ for any α and β in LM_T†. (In particular, note that although negation does not

---

† This result follows from the definitions of connectives. For example:

$$h' \in St_h^{\alpha} \cup St_h^{\beta} \quad \text{iff } h' \in St_h \text{ and } \models_{h'}\alpha \text{ or } \models_{h'}\beta$$
$$\text{iff } h' \in St_h \text{ and } \models_{h'}\alpha \vee \beta, \text{ by the definition of } \vee$$
$$\text{iff } h' \in St_h^{\alpha \vee \beta}$$
$$\text{and so } St_h^{\alpha} \cup St_h^{\beta} = St_h^{\alpha \vee \beta}.$$

The others follow by analogous arguments.

correspond to set complement in this respect, logical denial does.) Hence, by definition of the connectives, F_{hM} is closed with respect to set operations and thus is a field. Indeed, if one accepts that countable disjunctions are wffs in the logic LM_T, then F_{hM} is closed with respect to countable union, and thus is a σ-algebra.

This result allows construction of a Kolmogorov probability space for the predictive conditional probabilities.

Definition
The *h–M probability space* of T is $(St_{hM}, F_{hM}, P_{hM})$.
The *predictive conditional probability* of wff α, given that M is performed on a system initially described by h, is $prob_h(\alpha : M) = P_{hM}(St_{hM}^\alpha)$.

Where M is a measurement of magnitude m, we may write $prob_h(\alpha : m)$ in place of $prob_h(\alpha : M)$. The probability space, which depends on h and M, is a space of states. It is generated from the successor states of h which might describe the system after measurement M. The probability assigned to a wff α is a measure of the subset of these successor states which find wff α true.

For example, consider the simplest case, where theory T has only a finite number of states. Here the probability assigned to wff α, given initial valuation h and measurement M, can be expressed as a weighted sum of the states in St_{hM}^α. If we let $St_{hM}^\alpha = \{h_i : 1 \leqslant i \leqslant n\}$, then $prob_h(\alpha : M)$ $= \sum_{i=1}^n w_{hM}(h_i)$. The weight factors w_{hM} are normalised to set St_{hM} and obey the usual conditions, i.e. $w_{hM}(h_i) \geqslant 0$ for each h_i, and $\sum_{k=1}^m w_{hM}(h_k) = 1$ for $St_{hM} = \{h_k : 1 \leqslant k \leqslant m\}$.

According to the present view, predictive conditionals are the fundamental probabilities of mechanical theories, and so the h–M spaces are the fundamental probability spaces. Thus the traditional analysis of classical mechanics, founded on a set X of events and a corresponding event space (X, F, P), is rejected. We shall see that this analysis can be derived from the definitions above in the special case of classical theories, but it cannot be generalised in non-classical cases. In general, probabilities are indeed probability measures defined over a Kolmogorov probability space, but this is a space of states depending on the initial valuation describing a system and on the measurement which is performed. Generally a theory requires many such probability spaces to generate its predictive conditional probabilities.

Thus the traditional relation between probabilities and logic is rejected. The σ-algebras in the probability spaces of a theory do not represent the logic of a theory or even a 'logic of states'. We saw above that a correspondence does exist between set operations in the σ-algebra and some of the connectives in the logic of a theory, but this algebra does not represent the logic LM_T. For one thing, it is denial rather than negation which is represented by the set complement.

It is possible to re-express the predictive conditional probabilities so that a wff gives the initial condition or so that the probability is assigned to a valuation of the theory. To do this, note the natural correspondence between wffs of a theory and its valuations (the first definition has already been given in chapter 4 above).

Definition
h_α is *characteristic for* α if $Th_\alpha = \{\beta : T \models \alpha \supset \beta\}$
β_h is *characteristic for h* if it is the conjunction of all propositions in ETh, the elementary truth-set of h.

h_α is thus the valuation which finds α true and all valid consequences of α in the theory true, and which makes no other truth-assignments. β_h is the conjunction of all simple propositions which are true in h. Using these terms, the fundamental predictive conditional probability can be re-expressed:

Definition

$$prob_\alpha(\beta : M) = prob_{h_\alpha}(\beta : M)$$
$$prob_{h'}(h : M) = prob_{h'}(\beta_h : M).$$

The first expression gives the probability of wff β given that M is performed on a system initially described by wff α. The second gives the probability that h' describes the system, given that M is performed on a system initially described by h. (One can now identify the weights in the earlier expression for a probability of a finite space by setting $w_{hM}(h')$ = $prob_h(h' : M)$.)

6.1.2 Some special cases

One can introduce probabilities where either the initial condition, or the measurement, is not specified. First one defines:

Definition
M_0 is the *trivial measurement* if every outcome of M_0 is valid in T.
The *trivial valuation* h_0 is such that $Th_0 = \{\alpha : T \models \alpha\}$.

The only outcomes of M_0 are valid wffs in the theory, and similarly h_0 only finds valid wffs to be true. One can now generate the following special cases of the probability function:

Definition

$$prob_h(\alpha) = prob_h(\alpha : M_0)$$
$$prob(\alpha : M) = prob_{h_0}(\alpha : M).$$

The first function is used when no particular measurement is specified, and the second when there is no initial condition. Corresponding to these are special probability spaces. For example, in the first case, the probability is defined over the space $(St_h, F_h, P_h) = (St_{hM_0}, F_{hM_0}, P_{hM_0})$. St_h contains all the successor states of h.

One can of course combine the definitions above in order to derive a probability in which no conditions at all are specified. The function is generated from an 'absolute', or unconditional, probability space (St, F, P), generated from the set $St = St_{h_0 M_0}$ containing all states in the theory.

Definition
The *absolute probability* of α is $prob(\alpha)$.

We shall see that this is particularly important in classical theories.

Definition
The *ratio* probability of α in β, given that M is performed on a system initially described by h, is

$$prob_h(\alpha/\beta: M) = \begin{cases} prob_h(\alpha \cdot \beta: M)/prob_h(\beta: M), & \text{if } prob_h(\beta: M) \neq 0 \\ 0 & \text{otherwise.} \end{cases}$$

This gives a measure of the states in which α is true in the subset of states which are M-successors of h and which find β true. It tells us the proportion of states in set St_{hM}^{β} in which α is true, for it follows from the definition that $prob_h(\alpha/\beta: M) = P_{hM}(St_{hM}^{\alpha} \cap St_{hM}^{\beta})/P_{hM}(St_{hM}^{\beta})$. Recall from the discussion in §5.1 above that functions of this form are themselves probability measures on the space $(St_{hM}, F_{hM}, P_{hM})$, and also over the probability space which is generated from the subset St_{hM}^{β} of St_{hM}.

Discussion of the predictive conditionals can be simplified using this definition, for one can show that any probability on a space $(St_{hM}, F_{hM}, P_{hM})$ can also be expressed as a ratio probability over the space (St_h, F_h, P_h), no matter what measurement M is specified.

Lemma 6.2
For any measurement M of T, h in H_T,

$$prob_h(\alpha: M) = prob_h(\alpha/\beta_M)$$

This follows from the fact that $St_{hM} = St_h^{\beta_M}$, which is a set in F_h, and so $St_{hM}^{\alpha} = St_h^{\alpha \cdot \beta_M}$, for any α in LM_T which is also in F_h. Hence $prob_h(\alpha: M) = P_{hM}(St_{hM}^{\alpha}) = P_h(St_h^{\alpha \cdot \beta_M})/P_h(St_h^{\beta_M}) = prob_h(\alpha/\beta_M)$.

Thus the probability spaces of fundamental interest in any theory T are the spaces (St_h, F_h, P_h), conditional on valuations of the theory. All predictive conditionals of form $prob_h(\alpha: M)$ can be generated from these. We shall see in the next section that in the special case of a classical theory T, the single 'absolute' probability space will generate all the assignments of the theory, and so this space (St, F, P) plays a special important role.

However, in general, in non-classical theories of mechanics, a family of probability spaces will be used. Different probability spaces are used to generate probabilities which have different initial conditions.

6.1.3 Other representations

We saw in chapter 5 that traditionally a probability space of propositions and a space of states is derived from the fundamental event space of a theory. Here, on the other hand, spaces of states are fundamental. One can, however, show that probability measures of states generate corresponding measures of wffs, of propositions or of values of observables. We shall also see in this section how the magnitudes can be represented by random variables on the conditional probability spaces, in a manner analogous to the traditional representation of observables.

Recall that each valuation of a theory, and hence also each state, is uniquely represented by its truth-set Th of wffs. It follows that we can associate with each of the conditional probability spaces of states (St_h, F_h, P_h) a corresponding space of truth-sets of wffs, (W_h, F_h, P_h), where Wh is the set of truth-sets of states in St_h, $W_h = \{Th' : h' \in St_h\}$. The field in this space corresponds exactly to the field in the space of states, being generated from subsets of the form $W_h^\alpha = \{Th' : h' \in St_h^\alpha\}$. Thus the probability measure over these sets of wffs exactly corresponds to the probability measure in the space of states. One should remark here that the set of truth-sets W_h is not made up from the entire set of wffs in the logic of the theory, but from a subset depending on the condition h. In fact, the wffs are all those which might be true in a successor state of h, i.e. $\cup_i Th_i : h_i \in St_h$.

In a similar way, one can associate with any space (St_h, F_h, P_h) of states a space (E_h, F_h, P_h) generated from sets of elementary propositions. In this case, E_h is the set $\{ETh' : h' \in St_h\}$ of elementary truth-sets of states in St_h. One can remark again that this union of elementary truth-sets will not contain all elementary propositions of a theory, but rather a proper subset, depending on h. In fact, it can be characterised as the set $E_T - EF_h$, since it is only the propositions which are false in h which cannot be true in a successor state. From this set one can derive a kind of conditional analogue of the event space of classical theories. Recall that every elementary proposition is of the form (m, Δ) for m a magnitude in M_T and Δ a Borel subset of V_m. Thus corresponding to each set E_h will be another set X_h of values occurring in these propositions.

We saw earlier that in classical theories one single probability space (St, F, P) will generate all probabilities of the theory. It follows that a single space of values (X, F, P) can be derived from it. Set X now contains all values of all magnitudes in the theory, and so represents the set of events which is fundamental to the traditional analysis. This set generates the familiar phase space of a classical theory. However, where theories are non-classical, and so use more than one conditional space to generate their probabilities, there will be no such single space of 'events'. Instead, one can only associate with each conditional space of states another conditional space of values, or range of values, of the magnitudes. Just as the conditional space of wffs was generated from a proper subset of wffs which depended on the condition, so the conditional 'event space' (X_h, F_h, P_h) will also depend on h. Only a subset of magnitudes and their values will be represented.

Although only conditional spaces are used in non-classical theories, it is still possible to use random variables to represent the magnitudes of a theory. In fact, each magnitude must be represented by a family of probability distributions, defined on each of the conditional probability spaces. We shall now discuss this representation on the fundamental conditional spaces of states.

First, a real-valued function f_m is associated with any magnitude m of a theory. This is defined as follows:

Definition
f_m is the *function representing magnitude* m of theory T if $f_m : \text{St} \to \mathbb{R}$, such that:

$$f_m(h') = \begin{cases} r_i, & \text{where } \models_{h'}(m, r_i) \\ 0 & \text{otherwise} \end{cases}$$

f_m takes each state of theory T to the value found true of m in that state. If no atomic m-proposition is true in the state, then f_m takes value 0. The definition of a valuation ensures that function f_m is always well defined. For instance, no state h can be assigned to more than one value, since no state can find more than one atomic m-proposition true.

The function f_m will generate random variables on each probability space of a theory.

Definition
The *random variable* f_{m_h}, *representing* m on the probability space (St_h, F_h, P_h), is the restriction of f_m to the subset St_{hM} of St.

$f_{m_{hM}}$ coincides with f_m, but is defined only over domain St_{hM}. Similarly, one can define the random variables f_{m_h} on each space (St_h, F_h, P_h) as restrictions of f_m to domain St_h. Using this definition, one can associate with each magnitude of a theory a probability distribution corresponding to each probability space of the theory. These are defined in the usual way:

Definition
The *probability distribution of* m on space $(\text{St}_{hM}, F_{hM}, P_{hM})$ is the function $P^m(B) = P_{hM}(f_{m_{hM}}^{-1}(B))$.

This tells us how 'likely' it is, given that M is performed on a system initially described by h, that the value of m lies in the Borel set B. Again, an analogous condition applied to f_{m_h} will define the probability distribution of m on the space (St_h, F_h, P_h).

An expected value for each magnitude can also be derived in the traditional way:

Definition
The *expected value of magnitude m*, given initial h and measurement M, is
$\langle (m: M) \rangle_h = \langle (f_{m_{hM}}) \rangle$.

This is just the expected value of the corresponding random variable (see §5.1). It gives a kind of 'average' of the values found true of m in the states of St_{hM}. It is also possible to derive an expression for the dispersion, or 'uncertainty', of magnitude m, given h and M (see §5.1):

Definition
$$\Delta_{hM} m = \Delta f_{m_{hM}}.$$

Similar definitions will introduce the expected value and uncertainty of m on the spaces (St_h, F_h, P_h) of a theory.

Thus once we recognise the conditional nature of probabilities in mechanical theories, much of the traditional analysis in terms of random variables and associated concepts can be applied generally to any theory of mechanics.

6.1.4 Some properties of the probabilities

In the next section, we see that in classical theories a single probability space generates all the probability assignments, and so there is a traditional 'event space' in this special case. However, in general, where many different spaces are needed for a theory, probabilities have pecularities similar to those which are evident in quantum mechanics. Some of these peculiarities will now be investigated.

First, a principle analogous to von Neumann's Projection Postulate holds in any theory of mechanics. Recall that according to this principle (von Neumann 1955),

> ... if a system is initially found in a state in which the values of A cannot be predicted with certainty, then this state is transformed by a measurement of A ... into another state: namely into one in which the value of A is uniquely determined.

A similar rule can be derived from the analysis above, and is expressed in the following lemma.

Lemma 6.3
If a measurement of m, M has outcome p, then the system initially

described by h, will after M be described by h_M which is such that $prob_{h_M}(\mathrm{p}:m) = 1$.

To prove this, recall that $prob_h(\mathrm{p}:M) = P_{hM}(\mathrm{St}^{\mathrm{p}}_{hM})$. Clearly $\mathrm{St}^{\mathrm{p}}_{hM} \subseteq \mathrm{St}_{hM}$. To show that $\mathrm{St}_{hM} \subseteq \mathrm{St}^{\mathrm{p}}_{hM}$, let $\mathrm{p} = (m, \Delta)$. Since M is a measurement of m, some m-proposition $\mathrm{q} = (m, \Delta')$ is true in h_M, by definition a measurement of m. However, $h_m \mathrm{S} h$, and so by condition 1 defining S (page 81), q does not contradict p. It follows by condition 2 defining S, that for any $h*$ in $\mathrm{St}_{hM}, h_M \leqslant h*$, i.e. the truth-assignment to p is retained in any successor of h which might result from M. Thus $\mathrm{St}_{hM} = \mathrm{St}^{\mathrm{p}}_{hM}$, and so the lemma is proved.

Since there is no reason to suppose that the initial valuation h is also such that $prob_h(\mathrm{p}: M) = 1$, one must conclude from lemma 6.2 that the description of a system may be changed after a measurement M. In any theory, the valuation describing a system may be altered in a radical but well-defined way, exactly as we find in quantum theories. A version of von Neumann's Projection Postulate is found in every non-classical theory.

One can also show that, in general, theories employ a principle analogous to the Uncertainty Principle of quantum mechanics. Uncertainty relations exist for incompatible magnitudes of any non-classical theory. These can be stated as follows for arbitrary magnitudes m and n:

Lemma 6.4

If m, n are absolutely incompatible, then for any atomic m-proposition p, and valuation h in H_T, as $prob_h(\mathrm{p}) \to 1$, $prob_h(\mathrm{q}) \to 0$, for any atomic n-proposition q.

As the probability assigned to a precise value of m approaches 1, the probability assigned to any precise value of n approaches 0. Recall that since magnitudes m and n are absolutely incompatible, no state can decide all the m- and n-propositions, so as $prob_h(\mathrm{p}) \to 1$, then $\mathrm{St}^{\mathrm{p}}_h \to \mathrm{St}_h$, by definition of the probabilities. However, since m and n are incompatible, $\mathrm{St}^{\mathrm{p}}_h \cap \mathrm{St}^{\mathrm{q}}_h = \emptyset$, so as $\mathrm{St}^{\mathrm{p}}_h + \mathrm{St}_h$, $\mathrm{St}^{\mathrm{q}}_h \to \emptyset$, and $prob_h(\mathrm{q}) \to 0$. No valuation can predict an atomic value for each magnitude with certainty.

This principle can also be expressed in terms of random variables and their probability distributions. Recall that each magnitude m is associated with function f_m, and hence with random variables f_{m_h} on each probability space of the theory. In this way a probability distribution P_h is associated with each magnitude and each probability space. Now, as set $\mathrm{St}^{\mathrm{p}}_h \to \mathrm{St}_h$, $f_{m_h} \to \langle (m:M) \rangle_h$ and so $\Delta_h m \to 0$. However, since in this case $\mathrm{St}^{\mathrm{q}}_h \to \emptyset$, f_{n_h} increasingly diverges from $\langle (n:N) \rangle_h$ and so the uncertainty of this magnitude increases away from 0. Where m and n are incompatible, it is impossible that for some h, $\mathrm{St}^{\mathrm{p}}_h = \mathrm{St}^{\mathrm{q}}_h = \mathrm{St}_h$, and so it is impossible for the uncertainty of both magnitudes to be 0 in a single state.

It is also possible to show that, in general, inequalities which hold of classical probabilities, such as Bell's Inequality, fail in non-classical theories. We shall not discuss here the detail of Bell's argument, which concerned the spin $\frac{1}{2}$ experiment already described in the discussion of the EPR paradox in chapter 1 above[†]. However, the general principle involved, which ensures that such inequalities do not usually hold in non-classical theories, is very simple and well worth remarking on. Recall from the discussion of logic that the 'Law' of Excluded Middle does not hold in logic LM, i.e. the disjunction $\alpha \vee \neg\alpha$ is not universally valid (see lemma 3.6, §3.1.5). In non-classical theories this does not even hold in the logic of states—a non-classical state can find a wff α undecided, in which case $\alpha \vee \neg\alpha$ fails to be true even in this state. It follows that the wff in lemma 3.7, §3.1.5, will not hold even in the states of a non-classical theory, i.e. wff β is not generally equivalent to a conjunction $\beta \cdot (\alpha \vee \neg\alpha)$, and so is not equivalent to the disjunction $(\beta \cdot \alpha) \vee (\beta \cdot \neg\alpha)$. The probability of β will not therefore be the same as the probability of this disjunction.

Consider, for example, the space $(\mathrm{St}_h, F_h, P_h)$. The probability according to h of some wff β is a probability measure of the set St_h^β, the set of successor-states of h finding β true, according to the earlier definition. The probability of the disjunction $(\beta \cdot \alpha) \vee (\beta \cdot \neg\alpha)$ is similarly a measure of the union of sets $\mathrm{St}_h^{\beta \cdot \alpha} \cup \mathrm{St}_h^{\beta \cdot \neg\alpha}$. However, this union of sets may not contain all members of St_h^β, for there may be a non-classical state h' in St_h which finds β true, and hence $h' \in \mathrm{St}_h^\beta$, but which finds α undecided, and hence $h' \notin \mathrm{St}_h^{\beta \cdot \alpha}$ and $h' \notin \mathrm{St}_h^{\beta \cdot \neg\alpha}$. Thus $\mathrm{St}_h^\beta \neq \mathrm{St}_h^{\beta \cdot \alpha} \cup \mathrm{St}_h^{\beta \cdot \neg\alpha}$, and so it follows that

Lemma 6.5

$$prob_h(\beta) \neq prob_h(\beta \cdot \alpha) + prob(\beta \cdot \neg\alpha).$$

The failure of this equality leads to the failure of more complex expressions, such as Bell's Inequality, for these inequalities are derived from it.

Another property of the predictive conditional probabilities is that they are measurement dependent. The probability assigned by any valuation to a wff will vary according to the measurement which is specified. In general:

Lemma 6.6

$$prob_h(\alpha : m) \neq prob_h(\alpha : n).$$

[†] See Bell (1964). Discussions of Bell's contributions and references to other papers can be found, for example, in Jammer (1974) or Bub (1974). There is now experimental support for the failure of Bell's Inequality in quantum mechanics; see, for example, Accardi (1982) or Wightman (1981).

The probability that wff α is true given that a measurement of m is performed on a system initially described by h differs in general from the probability assigned to the same wff given that n is measured on the same initial system. This result follows simply from the fact that $St_{hm} \neq St_{hn}$ when the magnitudes concerned are not compatible. Clearly, when m and n *are* absolutely compatible, then there is no such measurement dependence, for in this case $St_{hm} = St_{hn}$.

Finally, one can show that according to this analysis of probabilities, sequential probabilities are calculated by means of a rule which clearly resembles the rule used in quantum mechanics (see the discussion in §6.3.4 below). We shall see in the next chapter that in the special case of classical theories, joint probabilities are used to calculate the probability of a sequence, but in general, sequential and joint probabilities do not coincide.

Consider, for example, the simple case of a sequence with just two members, where measurements at each stage are unspecified. The expression $prob_h\langle \alpha, \beta \rangle$ gives the probability of first finding α true after some measurement on the system initially described by h, and next finding β true after some other measurement has been performed, assuming that the system does not evolve between measurements. It is natural to assume that this is calculated as the probability of finding α given initial h multiplied by the probability of finding β given that α was initially found true after h. This can be expressed:

$$prob_h\langle \alpha, \beta \rangle = prob_h(\alpha)prob_{h,\alpha}(\beta).$$

The second term on the right-hand side expresses the probability of β being true, given that α was previously found true on a system initially described by h. To calculate such a probability, one must consider all those states of the theory which are β-successors of some valuation which is itself an α-successor of h. That is to say, one considers first all the valuations which are successors of h and find α true, and then considers the successor states which find β true of each of these valuations. The probability $prob_{h,\alpha}(\beta)$ will thus be a measure, depending on h and α, of the union of sets $\cup_i St^{\beta}_{h_i}$, where $h_i Sh$ and $\models_{h_i} \alpha$. If we let the set S^{α}_h be the set of successor valuations of h finding α true, $S^{\alpha}_h = \{h' \in H_T : h' \, S \, h \text{ and} \models_{h'} \alpha\}$, then this union of successor-sets will be expressed as $\cup_i St^{\beta}_{h_i} : h_i \in S^{\alpha}_h$.

In general, where a sequence of n members is considered, one must take account of a whole series of such unions of successor-states, for in general

$$prob_h\langle \alpha_1, \ldots, \alpha_n \rangle = prob_h(\alpha_1)prob_{h,\alpha_1}(\alpha_2) \times \ldots \times prob_{h,\alpha_1\ldots\alpha_{n-1}}(\alpha_n)$$

where each term after the first on the right-hand side is computed by taking an appropriately weighted measure of a union of successor sets. We saw above that the second term is a measure of $\cup_i St^{\alpha_2}_{h_i} : h_i \in S^{\alpha_1}_h$, and by similar argument the third is $\cup_j St^{\alpha_3}_{h_j} : h_j \in S^{\alpha_2}_{h_i}$ and $h_i \in S^{\alpha_1}_h$, and so on for all subsequent terms. At each stage, we measure the set of states which are successors of valuations finding α_j true, which are themselves successors of valuations finding α_{j-1} true, and so on.

In the case where measurements are specified at each stage of the sequence, the same kind of rule is used for the computation, except that now one is concerned only with a subset of the successors at each stage. For example, in the two-element case, $prob_h \langle \alpha_1 : M_1, \alpha_2 : M_2 \rangle$ is computed by setting

$$prob_h \langle \alpha_1 : M_1, \alpha_2 : M_2 \rangle = prob_h (\alpha_1 : M_1) prob_{h, \alpha_1 : M_1} (\alpha_2 : M_2)$$

where the second term on the right-hand side is now a measure of those states in the theory which are in a set $St^{\alpha_2}_{h'M_2}$, where h' is in $S^{\alpha_1}_{hM_1}$. A similar modification to the expression above gives the general rule for the probability of a sequence of n members.

The analysis of sequential probabilities can also be expressed in terms of probability distributions, where the same kind of construction applies. In general, where magnitudes may be incompatible, a joint probability distribution does not give the probability of a sequence. Instead, one must take into account, at each stage of computing the probability, the fact that initial conditions may be changed. For example, the joint probability distribution $P^{m_1, \cdots m_n} (B_1, \ldots, B_n)$ expressing the probability that magnitude m_1 has a value in B_1, magnitude m_2 in B_2 and so on, is *not* also appropriate for calculating the probability that a measurement of m_1 will yield an outcome in B_1, and then a following measurement of m_2 a value in B_2, and so on. Instead, this sequential probability involves conditionalising on a series of new conditions. For instance, the probability of a two-member sequence $\langle B_1, B_2 \rangle$, associated with magnitudes m, n respectively, will be given by a product of the form $P^{fmh}(B_1) P^*(B_2)$. The first term is the familiar expression for the probability that a measurement of m yields an outcome in B_1, given initial condition h. The second term, however, involves a different distribution, generated by considering a weighted sum of all the different distributions $P^{fnh'}(B_2)$, for h' a successor of a valuation which is itself a successor of h, and which might result from a measurement of m which has an outcome in B_1.

In the following sections of this chapter, it will be argued that the classical use of joint probabilities for sequences, and the more complex expression used in quantum theories, can both be derived from the general analysis above. The properties of the successor relation in either theory determine the particular rule which is used.

6.1.5 *Summing up*

A new approach to probabilities in mechanics has been proposed. Each theory uses a family of probability spaces, (St_h, F_h, P_h), generated from the successor states of a valuation h. The fundamental probabilities $prob_h(\alpha : M)$ are predictive conditionals expressing the probability of wff α given that measurement M is performed on a system initially described by h. According to the fundamental definition, this probability is given by $P_h(\mathrm{St}^{\alpha}_{hM})$, so it is a regular Kolmogorov probability measure of the subset of successor states in St_h, which might result from M, and in which wff α is true.

Thus the basic definition is analogous to a derived result in the traditional analysis of mechanical theories, for we saw in chapter 5 that traditionally a space of states can be derived from a more fundamental 'event space', by means of Stone's Theorem. According to the present view, the analysis in terms of states is fundamental. Furthermore, in non-classical theories in general, there is no single probability space which can generate all assignments of a theory, and so there is no analogue of the traditional 'event space'. Instead, each theory uses a family of different probability spaces, each dependent on the initial condition of the probability.

In §6.1.4 some striking properties of probabilities in non-classical theories were discussed. It was shown that where there is no single probability space to generate the assignments of a theory, the probabilities have properties very similar to those which are evident in quantum mechanics. In particular, there is a principle analogous to von Neumann's Projection Postulate, according to which the state of a system may be altered drastically, but in a well-defined way, after measurement. There is also an analogue of the quantum Uncertainty Relations, according to which the values of incompatible magnitudes cannot be determined with arbitrary precision. The nearer a state comes to predicting the value of one magnitude with certainty, the less certain is any corresponding prediction about the other magnitude. It was also shown that probabilities in non-classical theories will be measurement dependent, in that the choice of measurement will affect the final state describing a system. Lastly we saw that sequential probabilities in non-classical theories must be calculated by use of a rule which is quite unlike the joint probabilities of classical theories.

So we see that, although probabilities are still defined according to Kolmogorov's principles, in non-classical theories they have properties strikingly similar to the properties evident in quantum mechanics. Where states are not always bivalent, more than one probability space must be associated with a theory, and the 'quantum' peculiarities arise.

6.2 The probabilities of classical mechanics

6.2.1 The absolute probability space

The fundamental probabilities of any theory are predictive conditionals, which are Kolmogorov probability measures over valuation-dependent probability spaces of states. Here we shall see that in classical theories, where by definition states are always bivalent, this new analysis largely coincides with the traditional account. It can be shown, for example, that in the special case of a classical theory, a single probability space is used, no analogue of the Projection Postulate is needed, there is no Uncertainty Principle, and sequential probabilities are calculated from corresponding joint probability spaces.

Recall that predictive conditional probabilities are defined over the probability spaces (St_h, F_h, P_h), depending on the initial valuation h. We saw that in any theory, measurement-dependent probabilities can be expressed as ratio probabilities on a space of this form (lemma 6.2 above). In classical theories one can go further, and show that any probability can be expressed as a ratio on the absolute probability space (St, F, P):

Lemma 6.7
$prob_h(\alpha) = prob(\alpha/\beta_h)$ in any classical theory.

Recall that in classical theories, succession coincides with containment (lemma 4.6). Where β_h is the characteristic wff for h, it follows that $St_h = St^{\beta_h}$, and so $St_h^\alpha = St^{\alpha \cdot \beta_h}$, hence $prob_h(\alpha) = P_h(St_h^\alpha) = P_h(St^{\alpha \cdot \beta_h}) = P(St^{\alpha \cdot \beta_h})/P(St_h) = prob(\alpha/\beta_h)$.

Thus in classical theories a single space of states (St, F, P) will generate all probabilities of the theory. In this special case, a predictive conditional probability coincides with a corresponding traditional 'conditional'. Recall that set St contains all states of the theory, F is the σ-algebra of sets of form St^α for α in LM_T and P is a probability measure over these sets of states.

We saw above, in §5.2, that an 'event space' (X, F, P), and a corresponding space of propositions, is traditionally associated with the probability space of states of a classical theory. Indeed, traditionally one derives the space of states by means of Stone's Theorem, from (X, F, P). Here it is possible to do the opposite. In a classical theory, where a single space of states is used by lemma 6.7 above, one can derive from (St, F, P) a corresponding 'event space' (X, F, P). As in the traditional analysis, one can regard the field F in the probability space of states (St, F, P) of a classical theory to be the Stone Space of the algebra of sets of propositions F_p in the space (X_p, F_p, P_p), and corresponding to this space one can define the 'event space' (X, F, P), where X is now the set of values of the magnitudes of T (cf discussion in §5.2.1). However, it must be stressed again that none of these algebras in the probability spaces of a theory represent that theory's logic (cf §4.1.3).

The fact that classical theories use a single probability space means that most of the traditional analysis of classical mechanics will still hold.

The only difference is that, where traditionally one takes the event space (X, F, P) as fundamental, here one takes the space (St, F, P). Thus, for example, one can represent classical magnitudes by a single random variable, but whereas traditionally this will be defined on set X, here they are functions from the set St.

Definition
The *random variable representing magnitude m* of a classical theory T is the function $f_m : St \to \mathbb{R}$, where

$$f_m(h') = r_i \quad \text{if} \quad \models_{h'}(m, r_i).$$

Since all states are bivalent, every h in St finds an atomic m-proposition true for every magnitude m of the theory, and so the random variables are well defined. Since only one probability space is needed in classical theories, a single random variable on the space (St, F, P) will represent each magnitude. An expected value and uncertainty (RMS deviation) are also associated with each classical magnitude, according to the usual definitions.

6.2.2 Some properties of classical probabilities

Probabilities in classical theories have some special properties. One can show that classical states correspond to bivalent probability assignments, and so states in these theories are not essentially statistical as they are in general. The classical version of the Projection Postulate is trivial, in the sense that after non-disturbing measurements, classical states are never changed. Similarly, there is no classical analogue of the Uncertainty Principle, since magnitudes are always compatible. Classical probabilities are not measurement-dependent, and the probability of a sequence in classical theories coincides with the corresponding joint probability.

First one can show that classical states generate bivalent probability assignments:

Lemma 6.8
For any state h of a classical theory T, and any α in LM_T, $prob_h(\alpha) = 1$ or $= 0$.

By definition, $prob_h(\alpha) = P_h(St_h^\alpha)$. However, h is a state, and classical states have no successors apart from themselves (see §4.2.3), and so h is the only member of St_h. If $\models_h \alpha$, then $St_h^\alpha = St_h$ and so $prob_h(\alpha) = 1$. If $\dashv_h \alpha$, then $St_h^\alpha = \emptyset$, and so $prob_h(\alpha) = 0$. Since there are no other possibilities, the lemma is established.

This special result has important consequences for our understanding of classical states, for these are both bivalent *truth-value* assignments to the wffs of a classical theory and also bivalent *probability* assignments to these wffs. The relation of truth-values to probabilities will be further discussed in chapter 7. However, one can see from this lemma that traditionally the truth-values t, f and probabilities 1 and 0 might be assumed to coincide. Certainly, this result shows that in the special case of classical theories, states are not fundamentally statistical. Only a special pair of probabilities, 1 and 0, is assigned.

According to the general version of von Neumann's Projection Postulate, given in lemma 6.3 above, the valuation describing a system after a measurement M has been performed is a valuation which predicts an outcome p of M with certainty. In general, this involves a change in the description of the system, since one cannot assume that the initial valuation also predicts this proposition with certainty.

In fact, a similar principle holds even in the traditional analysis of classical theories. A system may be described by a valuation h which is not a state, so that h generates non-bivalent probabilities. Such valuations are sometimes called 'probability states' of classical theories. When new information is produced by a measurement of the system, then a new description h_M is used which predicts the outcome of M with certainty. If we assume that the outcome of M was not known in h, then we clearly have a change in our description. A proposition which was not certain in h is now assigned probability 1 in h_M.

The fundamental difference between this principle in classical and non-classical theories, however, is that a classical *state* will never by changed after measurement. Although new truth-values may be assigned after measurements, in a classical theory we can assume these are consistent with earlier descriptions. The truth-values simply accumulate, and the state which finally describes reality is assumed to describe the earlier systems too. Thus the probabilities assigned by classical probability states can be understood simply in terms of ignorance. We use h to describe the system before M has been performed, while we are still ignorant of its outcome. After M we add new truth-values to those already established in h, to obtain the valuation h_M which is now used to describe the system. However, the new propositions found true in h_M are assumed to have been true even when h described the system. Similarly when a state is selected to describe the system, we suppose that this also described it earlier, when we did not know all the truth-values. Thus in classical theories we assume that the same classical state describes a system before and after any non-disturbing measurement. We have seen that in a non-classical theory we cannot make this assumption, and states may be changed after measurement.

The special properties of the classical state also ensure that there are no Uncertainty Relations, and also that the inequalities discussed earlier do hold in these theories. Both of these features arose in the general case because states can leave wffs undecided. In classical theories, where all states are bivalent, neither characteristic can arise. Each state decides each proposition and so no magnitudes are absolutely incompatible and hence there can be no Uncertainty Relations of the kind discussed earlier. Similarly, in the classical logic of states, Excluded Middle holds; i.e. wff $\alpha \vee \neg\alpha$ is true in every state, and so too is $\beta \equiv \beta \cdot (\alpha \vee \neg\alpha)$, for any α, β. It follows that $St^\beta = St^{\beta \cdot \alpha} \cup St^{\beta \cdot \neg\alpha}$, and so $prob(\beta) = prob(\beta \cdot \alpha) + prob(\beta \cdot \neg\alpha)$ in classical theories. Thus Bell's Inequality and other analogous relations will always hold in classical theories.

The special properties of classical states also ensure that in classical theories probabilities are not measurement-dependent. Given the same initial condition, the same probability will be assigned to a wff, no matter what magnitude we may choose to measure.

Lemma 6.9
In any classical theory T, for magnitudes m and n and any initial valuation h,

$$prob_h(\alpha : m) = prob_h(\alpha : n).$$

According to the fundamental definition, $prob_h(\alpha : M) = P_{hm}(St^\alpha_{hm}) = P_h(St^{\alpha \cdot \beta_m})$ by the definition of β_m, (in §6.1.1 above) and lemma 6.7 above. Similarly $prob_h(\alpha : N) = P_{hn}(St^{\alpha \cdot \beta_n}_h)$, but classical states are bivalent truth-value assignments, and so every state will find both β_m and β_n true. (Recall that these are disjunctions of all m- and all n-propositions respectively.) Thus in classical theories, $St^\alpha_h = St^{\alpha \cdot \beta_m}_h = St^{\alpha \cdot \beta_n}_h$, and so the two probabilities coincide.

Similar considerations show that in classical theories sequential and joint probabilities coincide. For example, consider the simple case of a two-member sequence. According to the general rule, discussed in §6.1.4 above, the probability of a two-element sequence is given by the expression

$$prob_h\langle \alpha, \beta \rangle = prob_h(\alpha)prob_{h\alpha}(\beta)$$

where the first term is, of course, a probability measure, depending on h, of the set St^α_h, and the second term is a weighted measure of the union of sets $\cup_i St^\beta_{h_i} : h_i \in S^\alpha_h$. In classical theories, where succession coincides with containment, this union is, in fact, simply the set $St^{\alpha \cdot \beta}_h$, $St^{\alpha \cdot \beta \cdot \beta_h}$, as one can easily show. Thus in classical theories the second term is simply a probability measure of this set, weighted by the probability assigned to St^α_h, and so $prob_{h\alpha}(\beta) = P_h(St^{\alpha \cdot \beta}_h)/P_h(St^\alpha_h)$. Thus in classical theories

$$prob_h\langle \alpha, \beta \rangle = P_h(St^\alpha_h)P_h(St^{\alpha \cdot \beta}_h)/P_h(St^\alpha_h)$$
$$= P_h(St^{\alpha \cdot \beta}_h)$$
$$= prob_h(\alpha \cdot \beta).$$

Similarly, in the case of an n-ary sequence, the union of successor-sets which is considered at each stage coincides in classical theories with the set of states finding all previous conditions true. The general expression for an n-ary sequence was

$$prob_h\langle\alpha_1,\ldots,\alpha_n\rangle = prob_h(\alpha_1)prob_{h,\alpha_1}(\alpha_2)\ldots prob_{h,\alpha_1,\ldots,\alpha_{n-1}}(\alpha_n).$$

According to earlier discussion, one is successively taking weighted measures of appropriate successor-sets and their unions, first St_h^α, then $\cup_i \mathrm{St}_{h_i}^{\alpha_2}:h_i\in S_h^{\alpha_1}$, $\cup_j \mathrm{St}_{h_j}^{\alpha_3}:h_j\in S_{h_i}^{\alpha_2}$ and $h_i\in S_h^{\alpha_1}$, and so on. However, in this case, where succession coincides with containment, $\cup_i \mathrm{St}_{h_i}^{\alpha_2}:h_i\in S_h^{\alpha_1} = \mathrm{St}_h^{\alpha_1\cdot\alpha_2}$, $\cup_j \mathrm{St}_{h_j}^{\alpha_3} = \mathrm{St}_h^{\alpha_1\cdot\alpha_2\cdot\alpha_3}$, and so on. At each stage one takes a measure of the successor-set of states associated with the conjunction of wffs, weighting it by the measure assigned to the previous set. Thus

Lemma 6.10

$$prob_h\langle\alpha_1,\ldots,\alpha_n\rangle = prob_h(\alpha_1\cdot\ldots\cdot\alpha_n).$$

This is easily seen, since $prob_h\langle\alpha_1,\ldots,\alpha_n\rangle = P_h(\mathrm{St}_h^{\alpha_1})P_h(\mathrm{St}_h^{\alpha_1\cdot\alpha_2})/P_h(\mathrm{St}_h^{\alpha_1})\ldots$ $P_h(\mathrm{St}_h^{\alpha_1\cdots\alpha_n})/P_h(\mathrm{St}_h^{\alpha_1\cdots\alpha_{n-1}}) = P_h(\mathrm{St}_h^{\alpha_1\cdots\alpha_n})$.

One can also derive this using the earlier result about ratio probabilities in classical theories. Consider again the case of a two-element sequence. In this case one can reason as follows:

$$
\begin{aligned}
prob\langle\alpha_1,\alpha_2\rangle &= prob(\alpha_1)prob_{\alpha_1}(\alpha_2) && \text{by the general rule}\\
&= prob(\alpha_1)prob(\alpha_2/\alpha_1) && \text{by lemma 6.8}\\
&= \frac{prob(\alpha_1)prob(\alpha_1\cdot\alpha_2)}{prob(\alpha_1)} && \begin{array}{l}\text{by the definition of}\\ \text{ratio probability}\end{array}\\
&= prob(\alpha_1\cdot\alpha_2).
\end{aligned}
$$

An inductive argument will establish the general case, that

$$
\begin{aligned}
prob\langle\alpha_1,\ldots,\alpha_n\rangle &= prob(\alpha_1)prob(\alpha_2/\alpha_1)prob(\alpha_3/\alpha_1\cdot\alpha_2)\ldots\\
&\quad \times prob(\alpha_n/\alpha_1\ldots\alpha_{n-1})\\
&= prob(\alpha_1\ldots\alpha_n).
\end{aligned}
$$

Since any probability can be expressed in terms of absolute probabilities (by lemma 6.7), this will establish the general result, that sequential probabilities do coincide with joint probabilities in classical theories.

6.2.3 Summing up

Probabilities in classical theories have special properties because of the special successor relation. Any predictive conditional can be re-expressed as a ratio probability on the absolute probability space, and so a single probability space generates all assignments of the theory.

From this it follows that the striking characteristics of the new probabilities do not arise in classical theories. Classical states generate bivalent probability assignments and so do not seem essentially statistical. The state of a system is never altered after an undisturbing measurement in a classical theory. Only trivial uncertainty relations arise, since magnitudes are always compatible. Probabilities do not depend on measurement, nor on the order in which measurements are performed on a system, for sequential probabilities coincide with joint probabilities.

In classical theories, this account of probabilities largely coincides with the traditional view. Where states are always bivalent and so a strong successor relation is used, the 'quantum' peculiarities do not arise.

6.3 The probabilities of quantum mechanics

6.3.1 Some general remarks

We have seen that in classical theories predictive conditional probabilities have many traditional properties, while in the general case the probabilities have some of the striking 'quantum' peculiarities. Here we discuss some of these peculiarities again in order to show that the analysis does indeed provide a logical foundation for quantum probabilities.

Recall first that two outstanding problems were encountered in the opening discussion of quantum theories, in §1.1.4. The first problem was the essentially statistical nature of quantum states. Even a state which is represented by an eigenvector of an observable operator, and so predicts a value of this observable with certainty, will generally make only statistical predictions about values of other magnitudes. The second problem concerned von Neumann's Projection Postulate. We saw that after measurement the state of a system may 'jump' in a way that is not described by the evolution equations of a theory. Both of these can now be resolved.

In the discussion of states in chapter 4, we saw that a state is essentially a valuation of the theory which is maximal with respect to simple truth-value assignments. The new analysis of probabilities in §6.1 shows how each state can be associated with a probability assignment. Classical states, which are bivalent valuations, were shown in §6.2 to have some special properties. In particular, they generate bivalent probability assignments which assign only the values 1 and 0 to wffs of the theory. Thus their assignment of truth-values, and of probabilities 1 and 0, appear to coincide. As a consequence, these probability assignments have traditionally seemed to characterise states, and it has therefore appeared paradoxical that states of quantum theories are not also bivalent in this way. However, on the present view we see that states are essentially statistical. They are maximal valuations and like all valuations they generate probability assignments according to the earlier definitions.

Only in the special classical case will they be bivalent valuations generating bivalent probability measures.

To understand that quantum states do correspond to regular Kolmogorov probabilities, one must recognise that they are essentially conditionals defined over probability spaces which depend on the initial description of a system. Once this is realised, the lack of a quantum analogue of classical phase space is understood. It is only in the special classical case, where states are bivalent, that all the conditional probabilities can be expressed as ratios of absolute probabilities generated from a single 'event space'. In theories which are not classical there will be no single probability space to generate all the probabilities, and so there is no single underlying phase space in these theories.

We can understand the nature of quantum states if we recall the general discussion of chapter 4. In non-classical theories such as quantum mechanics, states, which are essentially descriptions, are not perfect 'pictures' of reality. They cannot assign truth-values to all elementary descriptions, and so more than one quantum state may be used to describe the same unchanged reality. Thus the state describing a system may be altered, even after a non-disturbing measurement.

Naturally, the present view must not be confused with the 'ignorance' or 'ensemble' interpretation of quantum mechanics, already mentioned as discredited by the EPR experiment at least. According to this ensemble interpretation, quantum states simply fail to give sufficient information about the magnitudes of a theory: one assumes, for example, that a system 'has a real momentum value', and 'has a real position', and concludes that a quantum state may fail to tell us what these are, thus leaving us in straightforward ignorance. This analysis is rejected here. It is true that, according to the present view, quantum states must be carefully recognised as descriptions of reality and distinguished from the real states-of-affairs themselves. It is true also that, on this view, states are understood as 'weak' descriptions, related by a weak successor relation, which are unable to give an adequate 'picture' of reality. This, however, is not the same as saying that states are 'incomplete', in the sense that they fail to be maximal descriptions and leave us in straightforward ignorance of some truth-value.

For example, consider a quantum state h which finds the atomic position-proposition (q, r) to be true. We can therefore say that '(q, r) is true of the system described by h', or 'magnitude q has value r according to h'. This same state will, however, fail to find a corresponding momentum-proposition true, so that all propositions of form (p, r') are undecided in h. The ensemble interpretation mentioned above maintains that the system described by h 'really has a momentum value', although h fails to tell us what it is. However, according to the present view, while there is indeed a sense in which the state leaves us in ignorance—it cannot provide a bivalent description of this system—this is not straightforward 'ignorance of the momentum value'. We know from the structure of the theory that no state such as h could provide this kind of information, since momentum and position are incompatible magnitudes in this theory.

However, we must be very careful in our use of terms, and must reject metaphysical assertions like 'the system has a momentum value'. We can say that the magnitude position, or the magnitude momentum 'has a value' in some state, or for example that an atomic position-proposition is true in h, but this is not the same as saying that the real system itself 'has a momentum value'. It follows from the structure of the quantum theory that the magnitude momentum cannot have a value in h, or alternatively that no atomic momentum-proposition can be true in h. This is, however, a peculiarity of our description, i.e. of the magnitudes we are using to describe reality. It is not a property of that reality itself.

To make this distinction between theoretical terms and elements of reality is not to deny the obvious relation between a theory such as quantum mechanics and real physical measurement. Our magnitudes are clearly not an arbitrary or abstract choice, just as the theory itself is founded on observation and experiment rather than on quite abstract considerations. However, neither should this relation to observation obscure our dependence on theoretical terms and constructs. When we talk, for example, of 'position' in a quantum theory, or 'momentum', we certainly imply by these terms expectations about measurement, and quantum theory is certainly a good theory in so far as these expectations are realised. However, in using these terms we are not simply naming real physical properties or elements, and we should not admit all the extra metaphysical assumptions that such a claim implies.

Hence the ignorance evident in the use of a non-classical state such as h is not simple ignorance of a 'real momentum value'. The state is non-classical, which indicates that the choice of magnitudes of the theory is in some sense not adequate, for the structure of E_T is such that it cannot admit bivalent valuations. It is the very terms of quantum theory which leave us in ignorance. We admit that some kind of description is expressed when we say (for example) that 'the magnitude momentum has value r': for instance, it expresses some particular expectations of measurement. However, the description is not entirely adequate, and it is this inadequacy which makes it so important to distinguish our states, which are descriptions, from the states-of-affairs which they describe. We might use two different states, h which finds it true that 'momentum has value r', or h' which finds 'position has value r'', to describe the same physical state-of-affairs.

The general version of von Neumann's Projection Postulate was discussed in §6.1 above, and the classical analogue in §6.2. We saw that in any theory the outcome of a measurement is predicted with certainty after it has been performed. In non-classical theories this may mean that the state of a system is changed after measurement, but in classical theories no change in state will ever be involved. Once more, theories in general share the characteristics of quantum mechanics, while classical theories are an important special case. Once more, too, we properly understand the general situation only when we carefully distinguish states from physical states-of-affairs. The state describing a system may be altered by a non-disturbing measurement, and so probability assignments too will be altered, but this does not mean that the reality itself has changed (cf §4.2.3. This point will be discussed again in chapter 8).

Two fundamental problems of quantum probabilities are easily resolved if we accept this analysis of conditional probabilities. Furthermore, we saw that other problems too are resolved, for other peculiar features of quantum probabilities are also features of non-classical predictions. For example, probabilities vary according to the measurements performed on a system. Also, sequential probabilities must be calculated by a special rule which recognises the changes in condition at each stage of the sequence. Furthermore, where incompatible magnitudes are used in a theory, 'Uncertainty Principles' arise—the closer one comes to predicting a value of one magnitude with certainty, the less certain one is about values of incompatible magnitudes. Similarly, inequalities such as Bell's will fail in non-classical theories, as they do in quantum mechanics. All these characteristics were discussed in §6.1 above, as stemming from the non-classical nature of a theory. These properties of quantum predictions are explained if we accept this analysis as a logical foundation for probabilities in mechanics.

6.3.2 Some specific examples

To see in detail how this analysis of probabilities does explain some specific cases from quantum mechanics, one must first relate the logical terms used here to the Hilbert Space representation of quantum theories, outlined in chapter 1. Recall from that discussion that normed unit vectors of a Hilbert Space $\mathscr{H}_T$ represent the 'pure' states of quantum theory T. Quantum mechanical 'mixtures' are probability distributions over the pure states of a theory: they are therefore represented by weighted sums of the pure state vectors. Observables of the theory are represented by 'observable operators' in the Hilbert Space, and the spectrum or set of eigenvalues of these operators represents their possible measurement outcomes. Each value of a magnitude is thus associated with an eigenvector of its observable operator or, in the general continuous case, with a particular projection operator.

According to quantum mechanics, the probability assigned by the pure state ψ to value a_i of observable A depends on the magnitude of the component of vector ψ in the direction of the eigenvector α_i associated with eigenvalue a_i of operator $\hat{A}$, or more generally on the projection of vector ψ into the subspace $\mathscr{K}_{a_i}$ associated with a_i. Recall that in the simple case of a maximal observable A whose operator $\hat{A}$ has discrete distinct eigenvalues $a_1, \ldots, a_n$ and corresponding eigenvectors $\alpha_1, \ldots, \alpha_n$, we have the quantum probabilities

$$prob_\psi(a_i : A) = |(\alpha_i, \psi)|^2 = |P_{a_i}\psi|^2.$$

In the present terms, the pure quantum states represented by vectors of the Hilbert Space correspond to valuations of the quantum theory. In particular, the orthogonal 'rays' or atoms of the Hilbert Space $\mathscr{H}_T$ are characteristic valuations for the atomic propositions of the theory. For example, the eigenvector α_1 in the example above, which was associated with value a_1 of observable A, corresponds to the valuation $h_{(A,a_1)}$, characteristic for the atomic proposition (A, a_1). (To simplify the notation, we shall write h_{a_1} for $h_{(A,a_1)}$.) Quantum mechanical 'mixtures' are weighted sums of the pure quantum states, and hence correspond to weighted sums of characteristic valuations. They are not themselves necessarily associated with valuations of the theory, although they do generate probability assignments to the theory's propositions, just as valuations do.

We have seen that in quantum theories the inner product of two vectors is used to calculate conditional probabilities. Thus, according to the present view, this inner product is a measure of the corresponding successor-set of states. For example

$$|(\alpha_1, \psi)|^2 = P_\psi(St_\psi^{(A,a_1)}).$$

The quantum probability is therefore a measure of the set of successor states (i.e. of maximal valuations, not quantum 'states'†) which might result from a measurement of A and which find the atomic proposition (A, a_1) true.

If we do accept this analysis, it is possible, for example, to see why distinct values of an observable must be associated with eigenvectors which are orthogonal in the Hilbert Space. Two vectors are orthogonal when their mutual successor-sets of states are empty, and this will always be the case for distinct eigenvectors of an observable operator. That is:

Lemma 6.11

$$St_{ha_i}^{(A,a_j)} = 0 \quad \text{for } i \neq j.$$

There are no states which are successors of h_i, the characteristic valuation

---

† Quantum 'states' are generally valuations of the theory but are not necessarily maximal valuations or states in the logical sense used here.

for proposition (A, a_i), and which also find a different atomic proposition (A, a_j) to be true (by condition 2 defining succession; see discussion of lemma 6.3 above). It follows from this lemma that $P_{\alpha_i}(\text{St}_{\alpha_i}^{(A,a_j)}) = 0$, i.e. $prob_{\alpha_i}((A, a_j) : A) = 0$, as follows from the orthogonality of the vectors in quantum theory. In a similar way, one justifies the use of normed unit vectors to represent the pure states of a quantum theory, for

Lemma 6.12

$$\text{St}_{ha_i}^{(A,a_i)} = \text{St}_{ha_i}$$

and so $prob_{\alpha_i}((A, a_i) : A) = 1$, as follows in quantum theories from the fact that the corresponding vectors in Hilbert Space have unit norm.

The relations established above will allow some specific examples of quantum probability assignments to be examined. These are the examples already introduced in §5.3, as presenting problems for the traditional analysis of probabilities in mechanics.

First, consider Bub's example, involving a quantum theory which is represented by a three-dimensional Hilbert Space $\mathscr{H}_{\text{T}}$, and which uses incompatible magnitudes A and B, each associated with distinct discrete values a_1, a_2, a_3 and b_1, b_2, b_3 respectively, and corresponding eigenvectors $\alpha_1, \alpha_2, \alpha_3$ and $\beta_1, \beta_2, \beta_3$ (see §5.3.1). As remarked above, it seems puzzling that a quantum state which is represented by one eigenvector, say α_1, can assign non-zero probabilities to values of the incompatible observable. For example, $prob_{\alpha_1}(b_1 : B) = |(\beta_1, \alpha_1)|^2 \neq 0$, even though A and B are incompatible, and so we know that there is no state of the theory in which value a_1 is true of A, and also b_1 is true of B. As argued in §5.3, it follows from this kind of example that the quantum probabilities cannot be understood as traditional 'conditionals', i.e. ratio probabilities, which are derived from corresponding joint probability assignments. Since the joint probability must be zero, the corresponding ratio cannot be non-zero and so the quantum assignments are not traditional 'conditionals'.

The present analysis allows one to understand the quantum probabilities in a very simple way, for $prob_{\alpha_1}(b_1 : B)$ is now understood as the predictive conditional $prob_{ha_1}((B, b_1) : B)$, which gives the probability that proposition (B, b_1) is found true after B has been measured on a system initially described by proposition (A, a_1). This probability gives a measure of the set of successor states of h_{a_1}, the characteristic valuation for (A, a_1), which might result from a measurement of B and in which (B, b_1) is true. That is:

$$prob_{\alpha_1}(b_1 : B) = P_{ha_1}(\text{St}_{ha_1}^{(B,b_1)}).$$

Although the joint probability of propositions (A, a_1) and (B, b_1) must be zero, since A and B are incompatible, this does not mean that set $\text{St}_{ha_1}^{(B,b_1)}$

must be empty and so the conditional probability above will not generally be zero. Although no state finds both propositions true, a successor of a state in which (A, a_1) is true may find (B, b_1) true. In a non-classical theory such as this, the conditional probabilities may be non-zero even when the magnitudes are incompatible.

This example points out the important difference between traditional 'conditionals', i.e. the ratio probabilities, and predictive conditionals, as they have been defined above. The predictive probability is truly conditional in the sense that it is defined over a probability space which depends on the condition. Thus in the absolute space (St, F, P), the set which is measured for the joint probability is empty, $St^{(A,a_1)} \cap St^{(B,b_1)} = \varnothing$. However, the set measured for the conditional probability belongs to a different probability space, and this is non-empty: $St_{ha_1}^{(B,b_1)} \neq \varnothing$. Thus while the ratio probability $prob((B, b_1)/(A, a_1)) = 0$, the conditional probability $prob_{(A,a_1)}((B, b_1):B) \neq 0$. In non-classical theories, these probabilities are distinct.

The second example from quantum mechanics which caused problems for the traditional view concerned the measurement-dependence of quantum probabilities, especially in a sequence of predictions (see §5.3.1). We considered the case of three measurements M_A, M_B, M_A^*, performed so quickly that we assume no evolution of the system takes place. Here, magnitude A is measured first, then B and lastly A is measured again. It was noted that although value a_1 may be found by the first measurement, this value could not be predicted with certainty as the outcome of M_A^*. Yet we saw that when measurements of the same observables are performed in a different order the outcome may be predicted with certainty. When the sequence of measurements is M_B, M_A, M_A^*, so that B is measured first followed by two measurements of A, then we do predict that the outcome of M_A is certain after M_A^* also.

If the probabilities are understood as non-classical predictive conditionals, then this example too can be easily understood. In the first case we suppose that a_1 is first found true of A, then value b_1 (say) is found true of B after the second measurement. To predict the outcome of M_A^* one now conditionalises on the outcome of this preceding B-measurement, and so we measure the set $St_{hb_1,A}$ of states which are successors finding A-propositions true of valuations in which (B, b_1) is true. Clearly, we have no reason to suppose that every such state finds proposition (A, a_1) true, so $St_{hb_1,A} \neq St_{hb_1,A}^{(A,a_1)}$, and $prob_{hb_1}(A, a_1):A \neq 1$. However, in the case where the measurements are performed in the different order, the probability is conditional on different information, for in this case one first finds value b_1 (say) for B, then value a_1 after the measurement M_A. If one now wishes to predict the value of A after the second measurement M_A^* of this magnitude, one conditionalises on proposition (A, a_1) which was true in the preceding valuation, and so in this case one measures the set $St_{ha_1,A}$ of states which are successors finding A-propositions true of valuations in which (A, a_1) is true. As we have already established, in lemma 6.3, St_{ha_1A} = $St_{ha_1A}^{(Aa_1)}$, and so $prob_{ha_1}((A, a):A) = 1$. The outcome of M_A will be predicted with certainty for M_A^* when no measurement of B intervenes.

Once more it is the lack of bivalent states and the consequent weakness of the successor relation in non-classical theories which generates the quantum 'peculiarity'. States are not bivalent, so succession is not containment, and we must change the state we use to describe a reality after incompatible magnitudes are measured. When we predict the outcome of a measurement of A after a measurement of B, we conditionalise on a state which is different from the state used when a measurement of A is repeated. The intervening measurement of B alters the initial condition for the prediction, and thus alters the prediction itself.

6.3.3 Quantum sequential probabilities

In §6.1, sequential probabilities were discussed and we derived the general rule for their computation. In §6.2, it was shown that the classical use of joint probabilities to calculate the probabilities of sequences can be derived from this general rule. It can be derived because the classical successor relation is containment. In a similar way, other special properties of the successor relation will generate other special cases of this rule. We shall now discuss one particular kind of special rule, and will argue that this is used in quantum theories.

Recall that, in general, sequential probabilities are calculated according to the expression

$$prob_h\langle\alpha_1 : M_1, \ldots, \alpha_n : M_n\rangle = prob_h(\alpha_1 : M_1)prob_{h\alpha_1}(\alpha_2 : M_2) \times \ldots$$

$$\times prob_{h,\alpha_1\ldots,\alpha_{n-1}}(\alpha_n : M_n) \qquad (1)$$

where each term on the right-hand side is an appropriately weighted measure of a union of sets—the sets of states which are successors of appropriate preceding valuations. For instance, the second term is a measure of the union of sets $\cup_i St_{h_i}^{\alpha_2}$, where h_iSh and $\vDash_{h_i}\alpha$. Now, a special case of this general expression is the case where at each state of the calculation one measures not a union of successor-sets but some particular successor set. In such a case, at each stage of the computation one conditionalises on a particular valuation of the theory. Thus the rule has the following form:

$$prob_h\langle\alpha_1 : M_1, \ldots, \alpha_n : M_n\rangle = prob_h(\alpha_1 : M_1)prob_{h_1}(\alpha_2 : M_2) \times \ldots$$

$$\times prob_{h_{n-1}}(\alpha_n : M_n). \qquad (2)$$

Here the new valuation h_i, which is selected as the new initial condition at each stage of the calculation, depends on the previous initial valuation h_{i-1}, the measurement M_i and also on the wff α_i. Thus h_i must be a successor of h_{i-1} which might result from M_i and which is such that $\models_{h_i} \alpha_i$.

Clearly, the classical case is an instance of this expression, for in a classical theory, where succession is containment, one conditionalises at each stage on a valuation which contains all preceding conditions. That is, $h \leqslant h_1 \leqslant h_2 \leqslant \cdots \leqslant h_n$ in expression (2) above. Another different instance of this special case (2) of the general rule is the following:

$$prob_h \langle \alpha_1 : M_1, \ldots, \alpha_n : M_n \rangle = prob_h(\alpha_1 : M_1) prob_{\alpha_1}(\alpha_2 : M_2) \times \cdots$$

$$\times prob_{\alpha_{n-1}}(\alpha_n : M_n). \tag{3}$$

Here one conditionalises at each stage on the valuation which is characteristic for the new information. In a sense, one is losing earlier information about a system, so that at each stage of the computation only the immediately preceding valuation affects the new prediction. This rule clearly differs from that which is used in classical theories.

To see what rule is used in quantum theories, recall von Neumann's Projection Postulate, according to which the state of a system is altered after measurement in such a way that the outcome is predicted with certainty (see discussion in §1.1.4 and §6.1.4 above). Consider, for example, the case of an observable A, represented by operator $\hat{A}$ with distinct discrete eigenvalues. According to von Neumann, after a measurement of A which has outcome a, corresponding to eigenvector α of operator $\hat{A}$, the system is left in the state α (von Neumann (1955) p 214. See also §§1.1.4 and 6.1.4). According to this same principle, in a sequence of measurements of observables whose operators all have distinct discrete eigenvalues, one calculates the probability of a sequence as follows:

$$prob_\psi \langle a : A, b : B, \ldots, d : D \rangle = |(\alpha, \psi)|^2 |(\beta, \alpha)|^2 \cdots |(\delta, \gamma)|^2$$
$$= \|P_\alpha, \psi\|^2 \|P_\beta, \alpha\|^2 \cdots \|P_\delta, \gamma\|^2$$

where $\alpha, \beta, \gamma, \delta$ are the eigenvectors corresponding to values a, b, c, d, of observable operators $\hat{A}, \hat{B}, \hat{C}, \hat{D}$ respectively (where C occurs before D in the sequence). At each stage of the computation one 'picks out' a new pure state and conditionalises on it. Adopting the relations already discussed between the present logical terms and the Hilbert Space representation, one accepts this rule as an instance of (3). In abbreviated logical terms it is

$$prob_h \langle a : A, \ldots, d : D \rangle = prob_h(a : A) prob_{h_a}(b : B) \times$$
$$\cdots \times prob_{h_c}(d : D).$$

To calculate the probability that a measurement of A will yield outcome *a*, then a measurement of *B* finds outcome *b*, and so on, one conditionalises at each stage on the preceding outcome.

In the case where the operator representing a quantum observable has degenerate (i.e. non-distinct) or continuous eigenvalues, then it may seem that this kind of rule is not used, and instead a more general instance of (1) applies. Von Neumann himself argued that after a non-maximal measurement, for example one which has a degenerate eigenvalue as outcome, the state describing the system is afterwards unknown. In this case the outcome *a* corresponds not to an eigenvector α of operator $\hat{A}$, but to a subspace $\mathscr{H}_a$ of the Hilbert Space, which has dimension greater than 1. According to von Neumann, any vector in this subspace might be used after the measurement to describe the system (von Neumann 1955 pp 218ff). He therefore concludes that the reality is described by a quantum mechanical 'mixture'. The probabilities are assigned by projecting vectors into the whole subspace $\mathscr{H}_a$, and so by considering the magnitude of their components in some basis set of vectors for this subspace.

However, it has since been argued, and is now generally accepted, that von Neumann's analysis in such a case is incorrect. Lüders (1951) first proposed that von Neumann's expression be amended in such a way that after any measurement a pure quantum state is used to describe a system†. According to Lüders, if a system is initially described by ψ and is then subjected to a measurement of *A* which has degenerate eigenvalue *a* as outcome, then afterwards the system will be described by the vector which is the normalised projection of ψ into the subspace $\mathscr{H}_a$. According to Lüders, one conditionalises on this new pure state, which depends on the initial state ψ and on the outcome *a*.

According to von Neumann, a non-maximal measurement, given a pure initial state, results in a transition from a pure state to a mixture. According to Lüders (on the other hand) even non-maximal measurements yield pure states to describe the system. Thus, von Neumann's analysis corresponds to a very general version of the conditionalising rule (1) for mechanical theories, while Lüders' amendment to von Neumann suggests that in all cases the special version (3) is used. Since Lüders' amendment is generally accepted as correct, one concludes that (3) is appropriate in quantum theories. Both classical and quantum theories therefore use a version of (2), calculating a sequential probability by conditionalising at each stage on a new initial valuation.

We have seen in the discussion of classical theories that they use a strong successor relation, and certainly do not use rule (3) for calculating sequential probabilities. Succession coincides with containment, and so the information obtained in a sequence of measurements simply accumulates, and in fact the probability of a sequence is merely the

---

† For a discussion of this paper see, for example, Bub (1977, 1979).

corresponding joint probability. It is perhaps interesting to see what properties the successor relation must have in a theory to generate the rule (3), which is used in quantum theories. In fact, this rule will arise whenever the unions of successor-sets of states, which are measured at each stage in the general expression (1), coincide with the successor-sets of valuations characteristic of the outcomes of the measurements. This will happen when the valuations of a theory can always be 'linked' together by successors, in the following way:

Definition

H has the property of *linkage* if for any h in H_T and any state $h*$, there is an h' in H_T, such that $h*Sh'$, and $h'Sh$.

Any state can be 'derived' from any valuation, by an intermediary successor. Where the valuations of a theory have this property, then rule (3) above will be used. This property indicates the 'weakness' of a theory's successor relation, and contrasts sharply with the strong classical use of the containment relation.

6.3.4 Summing up

We have seen here that predictive conditional probabilities of non-classical theories have the peculiar features characterising quantum probabilities, so if quantum probabilities are supposed to be conditional assignments of a non-classical theory, many problems are resolved. We understand the essentially statistical nature of quantum states, as well as the 'jump' in state which is described by von Neumann's Projection Postulate. Specific examples, which raised problems for the traditional analysis of probabilities in mechanics, are analysed here and explained. As we saw in §6.1, other peculiarities, too, such as the measurement-dependence of probabilities, and the Uncertainty Principles, can also be explained. Lastly, in §3.3 we discussed in detail the quantum mechanical expression for calculating sequential probabilities, showing that this also can be regarded as a predictive conditionalising rule, appropriate to a non-classical theory.

All these points suggest that the present view of logic and probabilities may indeed provide a foundation for the logic and probabilities of quantum mechanics. Some wider implications of this view will be discussed in chapter 8 below, after a discussion in chapter 7 of the modal logic which expresses the predictions of a theory.

7 The Modal Logic of Predictions

7.1 Introduction to modal logic

7.1.1 *What is a modal logic?*

In chapter 2 we saw that a propositional logic uses logical connectives to combine simple propositions into complex expressions. For example, in classical bivalent logic (CBL), one can form conjunctions, disjunctions and implications from some fundamental elementary set of propositions or variables. In a three-valued system such as LM, one can also form such wffs as Tα for 'It is true that α', of Fα, 'It is false that α'. By definition, the logical connectives are truth-functional, i.e. the value assigned to a complex expression in any valuation depends on the values assigned to its constituents. In other words, the rules defining logical connectives can be expressed in a truth table (see §2.1.2).

Originally, a modal logic was supposed to be any propositional system which could express some sense of 'necessity' or 'possibility'. In some cases these were ordinary propositional systems, using only truth-functional connectives. For example, Lukasiewicz developed many-valued propositional logics, in an attempt to express modalities (see Borokowski 1970, McCall 1967). However, in the last few decades the term 'modal logic' has been used only for propositional systems which are augmented by at least one operator which is not truth-functional. It is by now accepted that the most interesting modalities need more complicated defining rules than those which can be expressed in a truth table. Here I shall describe what is currently the accepted analysis of modal operators, before introducing a more general and simpler analysis of my own. This will later be used to generate a modal logic for mechanics.

7.1.2 *'Possible Worlds'*

The traditional analysis of modal operators applies to classical bivalent systems (see, for example, Hughes and Cresswell 1968). The operators are added to classical bivalent logic, and generate what are known as the 'normal modal systems'. Here we shall first show how such a logic is defined, before discussing briefly (in the next section) some properties of these normal systems.

A normal modal logic is an extension of a classical bivalent system, and so includes its alphabet, rules of formation, and abbreviations. The alphabet of a modal extension MCBL of the abstract system CBL is simply extended by the introduction of a single primitive modal operator, $\square$†.

Alphabet of MCBL

p, q, r	propositional variables
$\neg$, $\supset$, $\square$	logical operators
(,)	brackets

Similarly, the Formation Rules for the modal logic are those of CBL, augmented by a rule for $\square$:

Formation Rules for MCBL

1 Any propositional variable p is a wff
2 If α, β are wffs, then so too are $\neg\alpha$, $\square\alpha$ and $\alpha \supset \beta$
3 Only strings generated from rules 1 and 2 are wffs. Normal conventions of bracketing apply.

The abbreviations of CBL are retained, and the following introduced:

Abbreviation

$$\lozenge\alpha = \sim \square \sim \alpha$$

Thus wffs of the modal logic include wffs of the propositional system, and modalised versions of these. For example, the following strings are wffs of a modal extension MCBL: $\neg(p \supset \square p)$, $\square(\alpha \vee \square\beta)$, $\square\neg\alpha \supset \alpha$, and so on.

In the next sections the operators $\square$ and $\lozenge$ will be used to express predictive modalities: $\square$ is used to express 'certainty', in the sense of a probability assignment of 1, and $\lozenge$ corresponds to a non-zero probability assignment. However, traditionally these operators are supposed to express some sense of 'necessity' or 'possibility'. Underlying the traditional analysis is the notion that 'necessary truth' can be understood as 'true in all possible worlds', while 'possible truth' corresponds to 'true in some possible world'. One need not accept this as a metaphysical insight to find it useful for defining modal operators in a logic.

Fundamental to the 'possible worlds' analysis of modal operators is the notion of a *frame* $\mathscr{F}$, which consists of a set W of points, or 'worlds', and a relation R over W‡. R is usually called an 'accessibility' relation. A model $\mathscr{M}$ based on $\mathscr{F}$, is an ordered triple $\mathscr{M} = \langle W, R, v \rangle$, where W, R are as

---

† Here we use the symbols $\square$ and $\lozenge$ for the 'necessity' and 'possibility' operators respectively. Another common pair of symbols is L and M.
‡ The 'possible worlds' analysis was developed by Kripke. The following discussion is based on a review by Cresswell (1975).

above, and v is a 'world valuation', a function which assigns propositional variables to subsets of W, $v(p) \subseteq W$. Intuitively, we understand v as picking out those 'possible worlds' of W in which proposition p is true. Valuation symbols $\models, \dashv$ which will be used in the valuation rules of the logic, are now three-place relation symbols, expressing the truth or falsity of a wff at a world, say $w \in W$, according to a model. For example, $\mathcal{M} \models_w \alpha$ is read 'α is true at w in the model $\mathcal{M}$', or alternatively 'α is true at w according to v'. The traditional definitions of operators in a modal extension of CBL are given in the following rules:

'Possible worlds' valuation rules for MCBL

Let $\mathcal{M} = \langle W, R, v \rangle$ be a model, $w \in W$.

1　Rule for propositional variables:
　　$\mathcal{M} \models_w p$ iff $w \in v(p)$, $\mathcal{M} \dashv_w p$ otherwise.
2　Rule for $\neg$:
　　$\mathcal{M} \models_w \neg\alpha$ iff $\mathcal{M} \dashv_w \alpha$, $\mathcal{M} \dashv_w \alpha$ otherwise.
3　Rule for $\supset$:
　　$\mathcal{M} \dashv_w \alpha \supset \beta$ iff $\mathcal{M} \models_w \alpha$ and $\mathcal{M} \dashv_w \beta$,
　　$\mathcal{M} \models_w \alpha \supset \beta$ otherwise.
4　Rule for $\square$:
　　$\mathcal{M} \models_w \square\alpha$ iff $\mathcal{M} \models_{w'} \alpha$ for all w': w'Rw,
　　$\mathcal{M} \dashv_w \square\alpha$ otherwise.

Since the logic is bivalent, these conditions do define the operators. Rule 1 asserts that proposition p is true at w in model $\mathcal{M}$ when w is in the set of 'worlds' assigned by v to p. (This means that $v(p)$ is indeed the set of worlds which find p true.) Rules 2 and 3 give the classical bivalent definitions for $\neg$ and $\supset$. Rule 4 defines the modal operator $\square$, by the condition that $\square\alpha$ is true at w in $\mathcal{M}$ only when α is true in all worlds 'accessible' to w in this model. The operator is bivalent, since $\square\alpha$ is false when this condition fails. Note that the rule defining $\square$ is the only one which refers to the accessibility relation R. Clearly this operator will vary according to the properties of this relation, so that different modal extensions will be generated from CBL, according to the different frames $\mathcal{F} = \langle W, R \rangle$ which are used.

The rule defining the 'possibility' operator $\Diamond$ can be derived from rule 4 above, together with the earlier definition. It is the following:

5　Rule for $\Diamond$:
　　$\mathcal{M} \models_w \Diamond\alpha$ iff $\mathcal{M} \models_{w'} \alpha$ for some w': w' R w,
　　$\mathcal{M} \dashv_w \Diamond\alpha$ otherwise.

$\Diamond\alpha$ is true at w in $\mathcal{M}$ when α is true in some world accessible to w in this model. $\Diamond\alpha$ is false when this condition fails. This operator too depends on the properties of relation R, and may be different in different modal extensions of CBL.

7.1.3 The normal modal systems

A normal modal logic is an extension MCBL of classical bivalent logic which can be generated by the conditions above. Different normal systems result from the use in these definitions of different frames. In particular, the properties of the accessibility relation R will be crucial to the characteristics of the normal modal system. A great deal of research has been done in the last ten or twenty years in analysing the relations between properties of R and properties of the resulting normal logic. (For a detailed bibliography on the subject see, for example, Hughes and Cresswell (1972).)

Before briefly stating some of the key results, one must first distinguish two different kinds of validity:

Definition
α is *valid in* $\mathcal{M}$ if $\mathcal{M} \models_w \alpha$ for every w in W.
α is *valid on* $\mathcal{F}$ if α is valid in every model based on $\mathcal{F}$.

In order to investigate the relation between properties of R and the resulting logic, we shall be concerned here with the validity of wffs on a frame.

The weakest normal modal system is the logic K, which is generated by a frame $\mathcal{F}$ in which R is assumed to have no special properties. The tautologies of K are wffs which are valid on any frame at all. It is a well-known result that K is characterised by the following axiom and Rule of Inference:

Rule of Necessitation (N):   If $\models \alpha$, then $\models \Box \alpha$.
Axiom K: $\Box(\alpha \supset \beta) \supset (\Box \alpha \supset \Box \beta)$.

One can easily check that the rule N holds in any modal extension MCBL, for if α is valid, it is true at every w in every model, and so whatever relation R is used it will be true also in all worlds accessible to w. Similarly, one can check that wff K is a tautology of any modal system MCBL, no matter what relation R is used. In fact it can be shown that the addition of this rule and axiom schema to an axiomatisation of CBL will generate all the tautologies of this weakest modal system as theorems.

By imposing additional conditions on R, we generate further conditions on the modal logic MCBL. For instance if R is supposed to be reflexive, then in addition to the axiom and rule above, the further axiom T will hold:

Axiom T: $\Box \alpha \supset \alpha$.

One can easily check that this is valid on any frame with a reflexive accessibility relation. The normal modal logic T, characterised by such frames, can be axiomatised by adding this axiom schema to the rule and axiom for K above, and to some set of axioms and rules for CBL.

The normal modal logic known as S4 results from the condition that R be not only reflexive, but also a transitive relation. This logic is characterised by the following axiom:

Axiom S4: $\Box\alpha \supset \Box\Box\alpha.$

The addition of this axiom to those for T gives an axiomatisation for S4. Where R is assumed to be symmetric, the 'Brouwerian' axiom B is valid:

Axiom B: $\alpha \supset \Box \Diamond\alpha.$

Finally, where R is an equivalence relation, i.e. reflexive, symmetric and transitive, then all the previous axioms hold, with the addition of the following:

Axiom S5: $\Diamond\alpha \supset \Box \Diamond\alpha.$

Many more extensions of CBL have been investigated, though these are perhaps the most well-known.

7.1.4 A more general analysis of modalities

The 'possible worlds' analysis, first developed by Kripke, has proved a fruitful method for defining interesting modal operators. However, it has been difficult to generalise and apply to other systems. For instance, it is not successfully applied to a logic which is not bivalent. Kripke (1975) and Segerberg (1965) have both attempted to give 'possible worlds' conditions for modal operators in a three-valued logic, but without great success—their method is very complex and requires several sets of worlds and accessibility relations. The 'possible worlds' analysis of modalities also meets problems when applied to predicate logics, for example, the problem of how individuals are identified in different worlds is still a matter for dispute. Here, a simpler and more general method for defining modalities will be introduced, and it is shown to generate an analogue of the 'possible worlds' conditions as a special case. This method will be used later to construct a modal logic of predictions for mechanics.

Instead of being based on a frame $\langle W, R \rangle$ the new analysis of modalities extends the propositional valuation relations and is based on the pair $\langle H, R \rangle$, where H is the set of valuations of the propositional logic and R is some relation over them. The valuation rules for variables and connectives in the propositional logic are retained, while the extra primitive operator of the modal extension defined as follows:

Rule for $\Box$:
$\models_h \Box\alpha$ iff $\models_{h'}\alpha$ for all h' in H such that $h'\,R\,h$.
$\dashv_h \Box\alpha$ otherwise.

Further modal operators can be derived from this. For example, one can introduce an operator defined by analogy with the normal modal operator $\Diamond$:

Rule for $\Diamond$:

$\models_h \Diamond\alpha$ iff $\models_{h'}\alpha$ for some h' in H such that $h'\,R\,h$.
$\dashv_h \Diamond\alpha$ otherwise.

$\Diamond\alpha$ holds in h if α is true in some valuation related to h.

Analogous weaker modal operators can be introduced into non-bivalent logics. Although we shall not be using them they can be described briefly as follows. One could, for example, use $\Box'$ to mean 'certainly will not be false', so that $\Box'\alpha$ holds in h when there is no related valuation in which α is false. Similarly $\Diamond'$ might mean 'is not impossible', in the sense that $\Diamond'\alpha$ would be true when α is not false in every related valuation.

Rule for $\Box'$:

$\models_h\Box'\alpha$ iff there is no $h': h'Rh$ and $\dashv_h\alpha$.
$\dashv_h\Box'\alpha$ otherwise.

Rule for $\Diamond'$:

$\models_h\Diamond'\alpha$ iff there is some $h': h'Rh$ and $\dashv_{h'}\alpha$.
$\dashv_h\Diamond'\alpha$ otherwise.

One can also use this kind of analysis for modal conditionals. For example, one could introduce the following sense of implication:

Rule for $>$:

$\models_h\alpha > \beta$ iff for all h' such that $\models_{h'}\alpha$ and $h'\,S\,h$, $\models_{h'}\beta$.
$\dashv_h\alpha > \beta$ otherwise.

Such conditionals have many interesting properties. Other similar connectives could also be defined. The structure $\langle H, R\rangle$ also provides a method for defining conditionals which is an alternative to the 'possible worlds' analysis which is generally used[†].

To show that the rules for $\Box$ and $\Diamond$ above do have the traditional normal operators as a special case, one can show that normal frames and models can be derived when there are only bivalent valuations. Let H be a set of bivalent valuations of a logic, and let R be a relation over H. From this structure $\langle H, R\rangle$ we can derive a normal frame $\langle W, R^*\rangle$ as follows. First one sets

$$W = \{Th : h \in H\}.$$

W is the set of truth-sets of the bivalent valuations. Relation R^* over W is

---

† In fact, one can show that this connective satisfies all but one of Stalnaker's modal axioms for the 'Stalnaker conditional'. Obviously, one can vary the definition in order to introduce conditionals with different properties (Stalnaker 1968).

derived from R as follows:

$$Th \, R^* \, Th' \qquad \text{iff } h \, R \, h'.$$

Two truth-sets, or 'worlds', are accessible when their corresponding valuations are related by R. A world-valuation function v is defined by setting

$$v(p) = \{Th : \models_h p\}.$$

v, a function from propositional variables to subsets of W, assigns each propositional variable to the set of truth-sets of valuations in which it is true. Using these definitions, one can generate a normal model $\mathcal{M}$ from the pair $\langle H, R \rangle$. To show that this model does generate equivalent conditions, we can show

Lemma 7.1

$$\models_h p \quad \text{iff } \mathcal{M} \models_w p.$$

This follows from the fact that H is assumed to contain only bivalent valuations, and so $\models_h p$ iff $h(p) = t$ iff $p \in Th$, also $\dashv_h p$ iff $p \notin Th$. Hence, by the earlier construction, $\mathcal{M} \models_w p$ iff $w \in v(p)$, and $\mathcal{M} \dashv_w p$ iff $w \notin v(p)$, as required. It follows from this result that all the traditional conditions can be derived from the analysis above in this special case of bivalent valuations.

Thus the new conditions will generate the normal modal extensions of classical bivalent logic. In the next section, we see that they can also be used to generate modal extensions of the three-valued logic LM which express probability statements of a mechanical theory.

7.2 A modal logic of mechanics

7.2.1 The modal extensions of LM_T

One can use the account of probabilities in chapter 6 and the analysis of modalities above to devise a modal extension of the logic of a theory. This will not only express the propositional descriptions of a theory, but will also express some of its probability statements. We have seen that the logic LM_T can make assertions about the truth, falsity or undecidedness of any wff in any valuation. In the modal extension MLM_T of this logic we shall also be able to express predictions about the outcome of measurements, for example asserting that a wff has probability 1 or 0 or has a non-zero probability after some measurement is performed.

The conditions defining the modal operators of MLM_T are derived from the analysis of probabilities, and so will be based on the pair $\langle H_T, S \rangle$ where H_T is the set of valuations of theory T and S is the successor relation of this theory. A 'certainty' operator $\square_M$ can be introduced for each measurement M, so that $\square_M$ corresponds to the statement that '$prob(\alpha:M) = 1$'. Derived from this will be a 'possibility' operator $\diamond_M$, corresponding to '$prob(\alpha:M) \neq 0$'. Before defining these operators, one first constructs the modal extension MLM$_T$ of LM$_T$ as follows:

Alphabet of MLM$_T$

p, q, r	propositional variables
$\square_M, \square_N, \ldots$	monadic operators, for measurements $M, N, \ldots$
$\neg, \supset$	logical connectives
(,)	bracket symbols

Since MLM$_T$ is an extension of LM$_T$, the propositional variables represent propositions in E$_T$ in accordance with earlier discussion (§3.1.1).

Formation Rules for MLM$_T$

1 Any propositional variable is a wff.
2 If α is a wff and M is a measurement of the theory, then $\square_M\alpha$ is a wff.
3 If α, β are wffs, then so are $\neg\alpha$ and $\alpha \supset \beta$.
4 Only strings generated by rules 1–3 are wffs. Normal conventions of bracketing apply.

The Greek letters $\alpha, \beta, \ldots$ now range over wffs of the modal extension, while p, q, r, $\ldots$ are retained for propositional variables. According to the rules above, all wffs of LM$_T$ will also be wffs of the modal extension. The modal wffs of particular interest are wffs of form $\square_M$p, i.e. $\square_m$p where p is an *m*-proposition, and M a measurement of *m*. $\square_m$p can be read 'p is certain after a measurement of *m*', or if p$= (m, \Delta)$, 'It is certain that a measurement of *m* will yield a value in Δ'. Examples of wffs in the logic are $\neg$p $\supset \square_M$(p $\supset$ q), $\square_M(\alpha \supset \square_N\beta)$ for measurements M and N.

The rules for evaluating propositional variables and logical connectives in MLM$_T$ are just those used in the logic LM$_T$, applied now to the extended set of formulae. For example, a proposition p is true in h if it is assigned to the element t in this valuation, and is false or undecided if it is assigned to f or to u respectively. The rules defining $\neg$ and $\supset$ are just those already given for LM$_T$, though they now apply also to modal formulae (see chapter 3). The rules defining the modal operators $\square_M$ are based on the pair $\langle$H$_T$, S$\rangle$, in accordance with the discussion of probabilities.

Rule for $\square_M$:

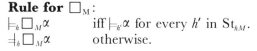

$\models_h \square_M\alpha$ ⠀⠀⠀⠀iff $\models_{h'}\alpha$ for every h' in St$_{hM}$.
$\nvdash_h \square_M\alpha$ ⠀⠀⠀⠀otherwise.

$\square_M\alpha$ is true in h when α is true in all successor states of h which might result from measurement M. This is indeed a 'certainty' operator, in the sense that

Lemma 7.2
$$\models_h \square_M \alpha \quad \text{iff } prob_h(\alpha : M) = 1$$

This follows simply from the definition of the probability and from the rule defining $\square_M$.

The abbreviations used in an extension MLM$_T$ include the abbreviations of LM$_T$. Logical connectives $\sim$, $\neg$, $\vee$, $\cdot$, $\supset$, $\equiv$, T, F, U, are all defined by the same abbreviations (see §3.1.4). In addition, one can introduce various modal operators, derived from $\square_M$. Several different senses of 'possibility' could be introduced. For example:

Abbreviation for MLM$_T$
(M-possibility) $\qquad \diamond_M \alpha = \sim \square_M \sim \alpha.$

One can show that this operator corresponds to the assignment of non-zero probability.

Lemma 7.3
$$\models_h \diamond_M \alpha \quad \text{iff } prob_h(\alpha : M) \neq 0.$$

This follows because the modal operator is characterised by the valuation rule that $\models_h \diamond_M \alpha$ iff $\models_h \alpha$ in some h' in St$_{hM}$, and $\nvDash_h \diamond_M \alpha$ otherwise, as one can derive from the rule for $\square_M$ and from the definition of $\diamond_M$. One can also introduce 'possibility' operators which correspond to the weaker condition that α is not false in every state in St$_{hM}$, as we briefly discussed in §7.1.4 above.

It is also possible to introduce 'absolute' modal operators, which are not measurement-dependent.

Abbreviations for MLM$_T$
$\square \alpha = \square_{M_0} \alpha$
$\diamond \alpha = \diamond_{M_0} \alpha$

These are dependent on the trivial measurement M_0, which has only wffs which are valid in the theory as outcomes (see §6.1.2). Clearly, for example, $\models_h \square \alpha$ iff $\models_{h'} \alpha$ for every h' in St$_h$, and so one can show that $\models_h \square \alpha$ iff $prob_h(\alpha) = 1$.

7.2.2 Modalities in mechanics

The logic MLM$_T$ of any theory T will have properties characterising the modalities of normal modal systems. For instance, the axiom and rule of the weakest normal system K also characterise any extension MLM$_T$:

Lemma 7.4
In any MLM$_T$, for any measurement M,

1   If $\models \alpha$, then $\models \Box_M \alpha$
2   $\models \Box_M (\alpha \supset \beta) \supset (\Box_M \alpha \supset \Box_M \beta)$

The rule of Necessitation (1) holds, and the κ axiom is valid. (1) is immediate. To show that (2) holds, suppose (i) $\models_h \Box_M (\alpha \supset \beta)$ and also (ii) $\models_h \Box_M \alpha$, for some h. By (i), $\models_{h'} \alpha \supset \beta$ for all h' in St_{hM}, and by (ii) $\models_{h'} \alpha$ for all h' in St_{hM}. Thus $\models_{h'} \beta$ by Modus Ponens, for all h' in St_{hM}, so by definition $\models_h \Box_M \beta$. It follows that the lemma is established. (Note that since the modal operators are bivalent, this case is all we need consider, for if either (i) or (ii) fails, the implication will be true.)

   Lemma 7.4 shows that the key properties of modal operators in normal modal systems, defined by the 'possible worlds' conditions, are also properties of the modal operators in the systems MLM$_T$. (Of course, these logics MLM$_T$ do not *contain* the normal modal logics, since they are not extensions of CBL, the classical bivalent propositional logic.) In fact, since the successor relation of any theory is reflexive, the axiom characterising the normal modal logic T also holds in any MLM$_T$. We show also that a restatement of this axiom in terms of the 'possibility' operators also holds.

Lemma 7.5
For any MLM$_T$, and measurement M:

1   $\models \Box_M \alpha \supset \alpha$
2   $\models T\alpha \supset \Diamond_M \alpha$

If a wff is certainly true in a valuation, then it will also be true. In a normal modal logic, this is often expressed by the equivalent wff '$\alpha \supset \Diamond \alpha$', which is also valid in the system T. However, this wff is not equivalent to (1) in the non-bivalent logic MLM$_T$, and in fact $\not\models \alpha \supset \Diamond \alpha$ in this logic, since α may be undecided and $\Diamond \alpha$ false. However, it is clear that (2) expresses the correct analogue of this normal wff in the system MLM$_T$. The normal formula corresponds to the intuitive requirement that 'what is true, is also possibly true', and this condition is expressed by (2), which is indeed a tautology of MLM$_T$.

   In fact, one can generate different modal extensions of the logic LM$_T$ according to the particular properties of S in a theory, much as one could generate different normal extensions of CBL. The properties of states in a theory determine the properties of succession, which in turn will determine the logic of predictions with different 'accessibility' relations used by a theory. As an example, let us just show that the following result will hold in any classical theory, where succession is the containment relation, although it will not be generally true.

Lemma 7.6

$\models Tp \equiv \Box p$ in MLM$_T$ for any classical T.

This follows from the fact that $\models Tp \supset \Box p$ in classical theories, by the special classical properties of S, which is the containment relation. Note that the converse of this wff is valid in any MLM$_T$, since S is reflexive, i.e. $\models \Box p \supset Tp$. In classical theories, where both wffs are valid, lemma 7.6 holds, and so for all elementary descriptions, truth and certainty coincide.

   We shall not develop any modal logic in detail here, since this will not be necessary for later discussion. In the next section this analysis of modalities will be used in our analysis of von Neumann's argument for a radically non-classical 'quantum logic'.

7.3 Von Neumann's 'Quantum logic'

In chapter 1 there was brief mention of a paper by Birkhoff and von Neumann (1975), in which they argued that quantum theories use a peculiarly 'quantum' logic. Their argument centres on a particular 'thought experiment', already mentioned in chapter 1. In fact, a more general argument, stemming from von Neumann's general analysis of properties and projection operators, has been recently revived and is currently popular. Here we shall look first at the quantum experiment described by Birkhoff and von Neumann, and later we shall examine the more important general arguments and the analysis of properties which underlies them. Finally, my own analysis of the 'calculus of properties' of a theory is presented in terms of the modalities discussed above.

7.3.1 The logical 'paradox'

In their paper *The Logic of Quantum Mechanics*, Birkhoff and von Neumann (1975) first suggest that the distributive identities may fail in the logic used by quantum theories. They then briefly cite the following experiment, as proof that distribution does actually fail. In their own words (Birkhoff and von Neumann 1975):

> That it [the distributive law] *does* break down [in quantum theories] is shown by the fact that if *a* denotes the experimental observation of a wave packet ψ on one side of a plane in ordinary space, a' correspondingly the observation of ψ on the other side, and *b* the observation of ψ in a state symmetric about the plane, then (as one can readily check):

$$b \cap (a \cup a') = b \cap \blacksquare = b > \odot = (b \cap a) = (b \cap a')$$
$$= (b \cap a) \cup (b \cap a')$$

(Here ■ and ⊙ are taken to represent the 'logically true' and 'logically false' propositions respectively.) This short passage is supposed by the authors to prove that quantum logic is non-distributive, and so one can only marvel at its ambiguity. Since publication of this paper, there has indeed been much argument and discussion about the passage (see e.g. Jammer 1974 pp 346ff). Here we shall look first at the example as it is presented in the paper, as an argument about propositions. At the end of this section we shall briefly review it again, re-expressed in terms of properties.

One can probably state the propositional argument from this example as follows. Proposition a concerns a position magnitude, A say, which has just two values in its value-set, corresponding to the observation of ψ on the left of some plane, and observation of ψ on the right. Thus we can form two A-propositions a, a', corresponding to these two possible values, which are relative complements. It follows that these are negations in the logic of the theory. Proposition b concerns a different, and incompatible magnitude B. Birkhoff and von Neumann use set operation and relation symbols to represent logical operations. Substituting proper logical terms, and stating the argument step by step, their analysis of the experiment can be re-expressed as follows:

$$1 \quad \models b \cdot (a \vee \neg a) \equiv b$$
$$2 \quad \models (b \cdot a) \equiv \odot$$
$$3 \quad \models (b \cdot \neg a) \equiv \odot$$
$$4 \quad \models (b \cdot a) \vee (b \cdot \neg a) \equiv \odot$$
$$5 \quad \not\models b \equiv \odot$$
$$6 \quad \not\models b \cdot (a \vee \neg a) \equiv (b \cdot a) \vee (b \cdot \neg a)$$

According to 1, the conjunction of b with the disjunction $(a \vee \neg a)$ is equivalent to the conjunction of b with any tautology. Clearly this assertion relies on the implicit assumption that the 'Law of Excluded Middle' holds. In particular it relies on the assumption that

$$1' \quad \models (a \vee \neg a) \equiv \blacksquare.$$

In addition it is assumed that

$$1'' \quad \models b \cdot \blacksquare \equiv b.$$

Thus from 1 and 1″ they conclude that

$$1^* \quad \models b \cdot (a \vee \neg a) \equiv b.$$

Lines 2 and 3 are supposed to follow from the construction of this particular experiment, in particular from the assumption that A and B are incompatible magnitudes, for it follows from their incompatibility that the conjunctions $(b \cdot a)$, $(b \cdot \neg a)$ can never be true, and since these authors assume bivalence in this 'quantum logic', they conclude that the conjunctions are in fact logically false and hence are equivalent to some logically false proposition. Line 4 follows from lines 2 and 3 and the definition of $\vee$. Line 5 is justified by the experiment, in particular from the choice of b as a proposition which is not logically false. The conclusion in line 6, that an instance of the distributive identities fails, is supposed to follow from lines 4 and 5.

Of course the authors have only shown by this argument that if one assumes that the Law of Excluded Middle holds, and hence that line 1 is valid, then distribution fails. They have given no reason for supposing that the distributive law does in fact break down. Indeed the authors themselves remark that distribution is 'the characteristic property of set combination', and so has been fundamental to our understanding of inference (Hooker 1975). Any logic without this property must be a radical departure from everyday logic. Yet they have not demonstrated that some other Boolean 'law', less fundamental to classical logic, might fail instead. In fact, a logic such as the non-bivalent system LM, which is distributive but only relatively orthocomplemented, will account for this kind of example while remaining essentially classical.

Thus according to the present view, the argument of Birkhoff and von Neumann fails at the outset. One does not accept the 'Law' of Excluded Middle as a logical truth, so line 1' cannot be assumed, and line 1 is not valid. Since $b \cdot (a \vee \neg a)$ is not logically equivalent to b, one cannot derive a failure of distribution from lines 4 and 5. This analysis of the experiment involves a much less radical departure from the accepted classical laws of logic, since it involves a slight generalisation of negation rather than a radical change in the properties of inference. Nothing in Birkhoff and von Neumann's paper refutes this kind of analysis.

7.3.2 *The general argument*

The underlying analysis of quantum theories which led von Neumann to conclude that quantum logic is not distributive was, in fact, a great deal more subtle than this example suggests, for his argument stemmed from certain properties of Hilbert Space. He argued that propositions in the logic of quantum descriptions corresponded to projection operators in the Hilbert Space of the quantum theory. This led him to set up a correspondence between propositions in the logic and subspaces of Hilbert Space, and between logical connectives and subspace operations in the Hilbert Space. These correspondences were supposed to establish that quantum logic is non-distributive.

Recall that any quantum theory has a set of magnitudes, i.e. observables, which are represented by 'observable' operators of the theory's Hilbert Space (see chapter 1 above). In von Neumann's terms magnitudes are quantities $\mathscr{R}$, $\mathscr{S}$, represented by operators R, S, i.e. his $\mathscr{R}$, $\mathscr{S}$ and R, S correspond to our R, S and $\hat{R}$, $\hat{S}$ respectively. Recall that the spectrum of an observable operator contains the values of its value-set.

According to von Neumann one can associate, with any quantity, properties of the system described by the quantum theory. Properties are characterised by the fact that they will be either verified or will fail to be verified, by an appropriate measurement of the system. For example, if a measurement of $\mathscr{R}$ has value λ as outcome, then von Neumann claims that the property '$\mathscr{R}$ has value λ' is verified. On the other hand, if any other value is the outcome, then this property fails to be verified. In von Neumann's own words (von Neumann 1955 p 249):

> Apart from the physical quantities $\mathscr{R}$, there exists another category of concepts that are important objects of physics—namely the properties of the states of the system S. Some such properties are: that a certain quantity $\mathscr{R}$ takes the value λ—or that the value of $\mathscr{R}$ is positive—or that the values of two simultaneously observable quantities $\mathscr{R}$, $\mathscr{S}$ are equal to λ and μ respectively—or that the sum of the squares of these values is > 1, etc.

Von Neumann then argues that one can associate, with any property, a special kind of quantity. These have only two values, 1 and 0, corresponding to whether we verify or fail to verify the associated property after a measurement. As von Neumann (1955) puts it:

> To each property $\mathscr{E}$ we can assign a quantity which we define as follows: each measurement which distinguished between the presence or absence of $\mathscr{E}$ is considered as a measurement of this quantity, such that its value is 1 if $\mathscr{E}$ is verified, and zero in the opposite case. This quantity which corresponds to $\mathscr{E}$ will also be denoted by $\mathscr{E}$.
>
> Such quantities take on only the values 0 and 1, and conversely each quantity $\mathscr{R}$ which is capable of these two values only, corresponds to a property $\mathscr{E}$ which is evidently this: 'the value of $\mathscr{R}$ is $\neq 0$'. The quantities $\mathscr{E}$ that correspond to the properties, are therefore characterised by this behavior.

Hence the correspondence between properties and quantities of this kind is one-to-one. Any quantity which has only two values represents a property, and any property is represented by a quantity of this kind.

Thus according to von Neumann, there is a special class of magnitudes in a quantum theory which represents the 'properties of the states of the system', and which are represented in Hilbert Space by a special class of observable operators. These operators have only the two eigenvalues 0 or 1, and so they are in fact the projection operators (von Neumann 1955):

That $\mathscr{E}$ takes on only the values 0, 1 can also be formulated as follows: Substituting $\mathscr{E}$ into the polynomial $F(\lambda) = \lambda - \lambda^2$ makes it vanish identically. If $\mathscr{E}$ has the operator E, then $F(\mathscr{E})$ has the operator $F(E) = E - E^2$, i.e. the condition is that $E - E^2 = 0$, or $E = E^2$. In other words: the operator E of $\mathscr{E}$ is a projection.

The projection operators represent magnitudes which have only two values. Von Neumann has thus established a unique correspondence between projection operators and 'properties'.

Recall now, however, that projection operators are themselves in one-to-one correspondence with the subspaces of Hilbert Space—each projection operator corresponds uniquely to that closed linear manifold which is its range (see §1.3.1). Von Neumann (1955) continues:

> The projections E therefore correspond to the properties $\mathscr{E}$ (through the agency of the corresponding quantities $\mathscr{E}$ which we just defined). If we introduce, along with the projections E, the closed linear manifolds $\mathscr{M}$ belonging to them $(E = P_{\mathscr{M}})$, then the closed linear manifolds correspond equally to the properties of $\mathscr{E}$.

In this way, von Neumann establishes a one-to-one correspondence between the properties $\mathscr{E}$ of a quantum system, the projection operators E in Hilbert Space which represent them and the subspaces of Hilbert Space: that is the closed linear manifolds $\mathscr{M}$ which are the ranges of the projection operators.

Having established this correspondence between properties, projections, and subspaces, von Neumann now proceeds to show that there is a relation between the 'certain presence' of a property in any state, and the containment of this state in the associated subspace. Von Neumann argues as follows:

> If in a state ϕ we want to determine whether or not a property $\mathscr{E}$ is verified, then we must measure the quantity $\mathscr{E}$, and ascertain whether its value is 1 or 0 (these processes are identical, by definition.) The probability of the former, i.e., that $\mathscr{E}$ is verified, is consequently equal to the expectation value of $\mathscr{E}$, i.e.,
>
> $$(E\phi, \phi) = \|E\phi\|^2 = \|P_{\mathscr{M}}\phi\|^2$$
>
> and that of the latter, i.e., that $\mathscr{E}$ is not verified, is equal to the expectation value of $1 - \mathscr{E}$, i.e.,
>
> $$((1 - E)\phi, \phi) = \|(1 - E)\phi\|^2 = \|\phi - P_{\mathscr{M}}\phi\|^2.$$

(The sum is of course equal to (ϕ, ϕ), i.e., to 1.) Consequently, $\mathscr{E}$ is certainly present or certainly absent, if the second or first probability respectively is equal to zero, i.e., for $P_{\mathscr{M}}\phi = \phi$ or $= 0$. That is, if ϕ belongs to $\mathcal{M}$ or is orthogonal to $\mathcal{M}$ respectively; or if ϕ belongs to $\mathcal{M}$ or to $\mathcal{H}_\infty - \mathcal{M}$.

$\mathcal{M}$ can therefore be defined as the set of all ϕ which possess the property $\mathscr{E}$ with certainty . . .

Here $\mathcal{H}_\infty$ is the Hilbert Space of the quantum theory. This correspondence between the 'certain presence' or 'certain absence' of a property and the inclusion or exclusion of a state in the corresponding subspace of Hilbert Space is fundamental to the claims of quantum logicians. They argue that the subspaces in fact 'represent' the properties and corresponding propositions, and that set relations among the subspaces represent the connectives of a quantum logic.

Consider, for example, the 'denial' of properties. In a passage following that quoted above von Neumann (1955) writes:

> If we call the property opposite that of $\mathscr{E}$ (the denial of $\mathscr{E}$) 'not $\mathscr{E}$', then it follows immediately from the above that if E, $\mathcal{M}$ belong to $\mathscr{E}$, then $1 - $ E, $\mathcal{H}_\infty - \mathcal{M}$ belong to 'not $\mathscr{E}$'.

Thus 'not $\mathscr{E}$' is defined so that the 'certain presence' of 'not $\mathscr{E}$', corresponds to the 'certain absence' of $\mathscr{E}$, and vice versa. By definition 'not $\mathscr{E}$' is the property assigned probability 1, if and only if property $\mathscr{E}$ is assigned probability 0. Clearly, as von Neumann argues, this sense of 'denial' is represented by the relations of E, $1 - $ E among the projection operators, or of $\mathcal{M}$, $\mathcal{H}_\infty - \mathcal{M}$ among subspaces of Hilbert Space.

The operations of 'conjunction' and 'disjunction' are defined by von Neumann so that where they are well defined they correspond to subspace operations in the Hilbert Space. For example, he defines 'conjunction' (von Neumann 1955):

> From properties $\mathscr{E}$, $\mathscr{F}$, which are simultaneously decidable, we can form the additional properties '$\mathscr{E}$ and $\mathscr{F}$' and '$\mathscr{E}$ or $\mathscr{F}$'. The quantity corresponding to '$\mathscr{E}$ and $\mathscr{F}$' is 1 if those corresponding to $\mathscr{E}$ and to $\mathscr{F}$ are both 1, and it is 0 if one of these is 0. Hence it is the product of these quantities. By [an earlier result], its operator is then the product of the operators of $\mathscr{E}$ and $\mathscr{F}$, i.e., EF. By [another earlier result], the corresponding closed linear manifold is the set $\mathscr{B}$ common to $\mathcal{M}$, $\mathcal{N}$.

'$\mathscr{E}$ or $\mathscr{F}$' is then defined as the property 'not $((\text{not } \mathscr{E}) \text{ and } (\text{not } \mathscr{F}))$', and von Neumann shows that this corresponds to the projection $\text{E} + \text{F} - \text{EF}$, which is in turn associated with the subspace $[\mathcal{M}, \mathcal{N}]$, known as the 'closure of $\mathcal{M}$ and $\mathcal{N}$', which is the smallest subspace of the Hilbert Space containing both $\mathcal{M}$ and $\mathcal{N}$. It follows that, according to von Neumann, the 'conjunction' of properties is represented by set intersection in the Hilbert Space, while their 'disjunction' is represented by the closure of their union. (Note that the closure [$\mathcal{M}$, $\mathcal{N}$] will generally be larger than the set union $\mathcal{M} \cup \mathcal{N}$, which is not generally itself a subspace of the Hilbert Space (von Neumann 1955).)

The 'simultaneous decidability' of properties, which von Neumann requires in his definition, is defined in terms of the simultaneous measurability of the associated quantities. If two quantities are not simultaneously measurable, then they cannot both be assigned a value 1 or 0 according to von Neumann's earlier definition, and so in such a case the 'connectives' above will not be well defined. There is no 'conjunction' or 'disjunction' of properties which are not simultaneously measurable. This leads von Neumann to conclude that there is a fundamental difference between classical and quantum 'logic' (von Neumann 1955):

> As can be seen, the relation between the properties of a physical system on the one hand, and the projections on the other, makes possible a sort of logical calculus with these. However, in contrast to the concepts of ordinary logic, this system is extended by the concept of 'simultaneous decidability' which is characteristic for quantum mechanics.

Clearly this 'sort of logical calculus' of properties is regarded as fundamentally a system of associated propositions. Indeed, in the very next sentence von Neumann (1955) refers explicitly to 'the calculus of these propositions, based on projections'. So the last step of his argument is to associate with each property a proposition, and to regard the 'connectives' he has defined among properties as essentially logical connectives of propositions. The conclusion that quantum logic is non-distributive is then established by showing that the corresponding subspace operations in Hilbert Space are not distributive (see, for example, Jauch 1968 p 27 or Shively 1980).

This is the general analysis of logic and Hilbert Space which underlies most modern arguments for a non-distributive 'quantum logic'.

7.3.3 Reply to von Neumann

In my view this argument breaks down in at least one important respect. One can accept von Neumann's analysis of properties as being distinct from magnitudes, and one can accept the relation he establishes between properties and projection operators. One can even accept the operations of 'denial', 'conjunction' and 'disjunction' which he defines among properties. However, he does not establish, and nor do the modern 'quantum logicians', that the resulting calculus is a propositional logic expressing the descriptions of a quantum theory. It is not a propositional logic in the sense that comparisons with classical logic suggest. Furthermore, we will see that the 'connectives' of von Neumann's calculus are actually partial but distributive operations.

One can agree that every atomic property, as these are defined by von Neumann, corresponds uniquely to a simple proposition of a quantum theory. To take his example, the property 'quantity $\mathscr{R}$ takes value λ' clearly corresponds to the proposition $(\mathscr{R}, \lambda)$. Similarly one can associate the property 'the value of $\mathscr{R}$ is positive' with the proposition $(\mathscr{R}, \Delta)$, where Δ is the subset of positive values in the value-set of $\mathscr{R}$. Likewise even complex properties, such as his example 'the values of two simultaneously measurable quantities $\mathscr{R}, \mathscr{S}$ are equal to λ and μ respectively', can be associated with the complex proposition $\alpha = (\mathrm{p} \cdot \mathrm{q})$, where p is $(\mathscr{R}, \lambda)$, and q is $(\mathscr{S}, \mu)$.

However, the fact that properties can be uniquely associated with wffs in this way does not mean that von Neumann's calculus of properties is a propositional logic, for propositions take truth-values and operations in a propositional logic must be defined in terms of these by valuation rules (see chapters 2 and 3 above). We find, however, from von Neumann's definitions that properties do not take truth-values, but instead are either verified or fail to be verified after they are measured. The operations of von Neumann's calculus are therefore founded not on truth-values, but on the 'certain presence' or 'certain absence' of a property after measurement. Although properties and propositions correspond, von Neumann's operations among properties are not logical connectives. They are not truth-conditional operations, expressed by valuation rules or by truth tables. The resulting calculus of properties is not therefore a propositional 'quantum logic'.

In fact, one can express von Neumann's calculus of properties in the modal logic of a theory which was discussed at the start of this chapter. We have seen already that a property may be associated with a formula α in the propositional logic $\mathrm{LM_T}$ of a theory. Indeed, one can also associate with each property a modal formula $\square_M \alpha$ in the modal extension $\mathrm{MLM_T}$, where M is a measurement of α. For example, suppose $\mathscr{E}$ is von Neumann's earlier example of a property, 'quantity $\mathscr{R}$ takes value λ'. We have seen that this corresponds to the proposition $\mathrm{p} = (\mathscr{R}, \lambda)$. However, $\mathscr{E}$ also corresponds to the modal formula $\square_{\mathscr{R}}\mathrm{p}$, expressing the proposition that 'quantity $\mathscr{R}$ is certain to take value λ after a measurement of $\mathscr{R}$', or $prob(\mathrm{p} : \mathscr{R}) = 1$ (see lemma 7.2). The truth of $\square_{\mathscr{R}}\mathrm{p}$ in a state corresponds to the 'certain presence' of the property $\mathscr{E}$ in that state, in the sense that value λ of $\mathscr{R}$ is predicted with certainty in this state.

The same analysis applies to complex properties. For example, the property mentioned above, that 'the two values of two simultaneously measurable quantities $\mathscr{R}, \mathscr{S}$ are equal to λ and μ respectively', corresponds to the wff $(\mathrm{p} \cdot \mathrm{q})$ in $\mathrm{LM_T}$. However, this property will also correspond to the modal wff $\square_M (\mathrm{p} \cdot \mathrm{q})$, where M is a measurement of both $\mathscr{R}$ and $\mathscr{S}$, guaranteed by the hypothesis that these are simultaneously measurable. This modal formula expresses the statement that 'the values λ of $\mathscr{R}$, and μ of $\mathscr{S}$, are predicted with certainty after measurement', and hence corresponds to the 'certain presence' of this complex property after it is measured.

One can therefore logically characterise a property as a special kind of wff in the modal extension $\mathrm{MLM_T}$ of a theory:

Definition
A *property* in theory T is a wff in logic $\mathrm{MLM_T}$ which has the form $\square_M \alpha$ for M a measurement of α.

The *atomic properties* in a theory are wffs of the form $\square_m \mathrm{p}$, where p is an *m*-proposition, and *m* indicates a measurement of this magnitude. Such a property corresponds to the statement that 'p is certain after a measurement of *m*', or $prob(\mathrm{p}{:}m) = 1$. In general, a property $\square_M \alpha$ corresponds to the proposition that 'wff α is certain to be found true if it is measured'. Properties are only well defined if the measurement M in $\square_M \alpha$ is indeed a measurement of α, in the sense that it has all the constituents of α as possible outcomes (see §4.2.1). If M is not a measurement of α then the modal formula $\square_M \alpha$ is not a property. For example, if α has constituent propositions p and q, which are incompatible, then there is no valuation deciding both p and q, and so there can be no measurement M which has both p and q as possible outcomes, and so there is no measurement M of α. Consequently $\square_M \alpha$ is not a property for any measurement M.

Using this definition, one can see at once the important difference between classical and quantum properties which is discussed by von Neumann, for in a classical theory all wffs are absolutely compatible and in this sense they are all 'simultaneously measurable'. Thus any wff in $\mathrm{LM_T}$ can be associated with a property, for whatever the constituent propositions in α one can always find a suitable measurement M to form a property $\square_M \alpha$. However, in non-classical theories such as quantum mechanics this will not generally be the case, as von Neumann has pointed out. Where wffs and magnitudes are incompatible it will be impossible to find a measurement which has all constituents of a wff as outcomes, and so there will be no property associated with the wff. In non-classical theories there are more wffs than properties.

In fact, one can express the operations of von Neumann's calculus in the modal logic $\mathrm{MLM_T}$ of a theory. Recall, for example, his definition of 'denial'. The property $\mathscr{E}$, associated with projection E and subspace $\mathscr{M}$, has as its 'denial' the property 'not $\mathscr{E}$', which corresponds to the projection $1 - \mathrm{E}$ and the subspace $\mathscr{H}_\infty - \mathscr{M}$. If $\mathscr{E}$ is 'certainly present' in a state ψ, then 'not $\mathscr{E}$' is 'certainly absent' in ψ, and vice versa (von Neumann 1955). It follows that these two properties correspond to the modal formulae $\square_M \alpha$ and $\square_M \sim \alpha$, for $\models_h \square_M \alpha$ iff $prob_h(\alpha : M) = 1$, and

$\models_h \Box_M \sim \alpha$ iff $prob_h(\alpha : M) = 0$. This suggests that we define the 'denial' of properties at the following operator in the modal logic MLM_T:

Definition

If $\Box_M \alpha$ is a property, then the *denial* of this is the property $\sim^* \Box_M \alpha = \Box_M \sim \alpha$.

We use the * to distinguish this operation $\sim^*$ among properties from the logical connective $\sim$ in the modal logic MLM_T.

Similarly, other operations in the calculus of properties can be defined in the logic MLM_T. The 'conjunction' and 'disjunction' of properties may be defined as follows:

Definition

If $\Box_M \alpha$ and $\Box_N \beta$ are properties corresponding to compatible wffs of LM_T, and M' is a measurement of both α and β, then

$$\text{('Conjunction'):} \quad \Box_M \alpha \cdot^* \Box_N \beta = \Box_{M'}(\alpha \cdot \beta)$$
$$\text{('Disjunction'):} \quad \Box_M \alpha \vee^* \Box_N \beta = \Box_{M'}(\alpha \vee \beta).$$

Again, $\cdot^*$ and $\vee^*$ are operations among properties. Since it is the case that $\models_h \Box_{M'}(\alpha \cdot \beta)$, iff $prob_h(\alpha : M) = 1$ and $prob_h(\beta : N) = 1$ (where M and N are measurements of α and β respectively), and $\dashv_h \Box_{M'}(\alpha \cdot \beta)$ when either of these wffs is not certain, one can suppose that the operation $\cdot^*$ corresponds to von Neumann's 'conjunction', defined in the passage quoted earlier. Similarly, $\vee^*$ does seem to correspond to his operation of 'disjunction'.

If P_T is the set of properties of theory T, then $\langle P_T, \cdot^*, \vee^*, \sim^* \rangle$ will be the 'calculus of properties' of this theory. One can see at once that there are important differences between the calculus of properties in a classical and a non-classical theory:

Lemma 7.7

If T is classical, then every wff in LM_T corresponds to a property in P_T, and the structure $\langle P_T, \cdot^*, \vee^*, \sim^* \rangle$ is an algebra.

If T is non-classical, then there are more wffs in LM_T than properties in P_T, and $\langle P_T, \cdot^*, \vee^*, \sim^* \rangle$ is a partial algebra.

If T is classical, then P_T exactly corresponds to the set of wffs in LM_T, since every wff can be associated with a property, as we discussed above. If T is non-classical this is not the case, and there will be wffs which do not have associated properties. In a classical theory too, all the operations are defined over all the properties in P_T, because all wffs are 'simultaneously measurable', and so $\langle P_T, \cdot^*, \vee^*, \sim^* \rangle$ is an algebra. However, in non-classical theories, $\cdot^*$ and $\vee^*$ are not defined among all properties of the theory, and so the calculus of properties $\langle P_T, \cdot^*, \vee^*, \sim^* \rangle$ is only a partial algebra.

The quantum experiment which was used by Birkhoff and von Neumann to argue for a non-distributive 'quantum logic' does indeed illustrate a difference between the classical and non-classical calculus of properties. The analysis in terms of propositions, given at the start of this section, can be re-expressed in terms of properties. Propositions a, $\neg a$ and b concerning magnitudes A and B are as before, and we use symbols A, B also to represent measurements of these magnitudes. The first line of the argument now reads as:

$$1 \quad \vDash \Box_B b \cdot^* (\Box_A a \vee {}^* \Box_A a) \equiv \Box_B b \cdot^* \blacksquare.$$

This is now justified by the assumption that

$$1' \quad \vDash \Box_A a \vee {}^* \Box_A \neg a$$

which is indeed a tautology (cf $1'$ in §7.3.1 above. Note that $\Box_A a \vee {}^* \Box_A \neg a = \Box_A (a \vee \neg a)$, which is valid because $a \vee \neg a$ is true in every state resulting from a measurement of A.) Von Neumann's example now proceeds by showing that the 'conjunctions' $(\Box_B b \cdot^* \Box_A a)$ and $(\Box_B b \cdot^* \Box_A \neg a)$ are not well defined, since by hypothesis the magnitudes A and B are not simultaneously measurable:

$$2 \quad (\Box_B b \cdot^* \Box_A a) = \odot$$
$$3 \quad (\Box_B b \cdot^* \Box_A \neg a) = \odot.$$

(Here symbol $\odot$ is taken to mean 'is ill formed', not 'is logically false', as it was in §7.3.1.) Thus the 'disjunction' of these properties is not well defined either:

$$4 \quad (\Box_B b \cdot^* \Box_A a) \vee {}^* (\Box_B b \cdot^* \Box_A \neg a) = \odot.$$

Since $\Box_B b$ is a well-defined property, however, it follows from 1 and 4 that the 'distributive' expressions are not identical:

$$5 \quad \Box_B b \neq \odot$$
$$6 \quad \Box_B b \cdot^* (\Box_A a \vee {}^* \Box_A \neg a)$$
$$\neq (\Box_B b \cdot^* \Box_A a) \vee {}^* (\Box_B b \cdot^* \Box_A \neg a).$$

This conclusion follows from the fact that the expression on the right-hand side is not well defined. Like the expressions in lines 2, 3 and 4, this is not a property.

Although von Neumann's example ostensibly shows a breakdown of distribution, closer analysis shows this is not the case. Distribution appears to fail in line 6, but in fact operations on the right-hand side are not defined. Where operations among properties are defined they are distributive, as one can easily show. This follows from the fact that the logical connectives are distributive†. Von Neumann's example has shown not that the operations among properties are non-distributive, but that they are not always defined in quantum theories. The example is indeed different from any classical case, since in a classical theory the operations would be defined everywhere, and so line 6 would indeed involve well-defined expressions and would give an instance of the distributive identity.

At any rate, the calculus of properties is not the logic of a theory. The operations among properties do not represent connectives of either the propositional logic LM or its modal extension. Hence the differences between the calculus of properties in a classical and a quantum theory are not differences in either the propositional or the modal logic of the theories. Our laws of reasoning are the same in any theory.

7.3.4 Summing up

Birkhoff and von Neumann describe an experiment which they claim shows that 'quantum logic' is non-distributive. We show in §7.3.1 that, as stated in their paper, their argument fails to establish this claim. One can conclude from their analysis only that some Boolean 'law of logic' fails in the quantum theory. There is no reason to suppose that it is distribution that fails. In fact, the logic LM, which is distributive but only relatively complemented, also accounts for their example and involves a much less radical departure from ordinary classical logic.

However, the argument underlying von Neumann's 'quantum logic' stems from his analysis of properties and their relation to projection operators in the Hilbert Space of a quantum theory. This general analysis is discussed in §7.3.2. Von Neumann argues that properties correspond to those special observables of a theory which have only two possible values. He therefore concludes that properties are represented by the projection operators in the Hilbert Space of a theory, and that they also correspond to the subspaces which are the ranges of these operators. Von Neumann then analyses operations among properties, which are defined in terms of

---

† For example, D2 (§2.2.1, p 42) is established by the following argument:

$$\Box_M \alpha \cdot^* (\Box_N \beta \vee^* \Box_L \gamma) = \mathrm{df}\Box_M \alpha \cdot^* (\Box_{N'}(\beta \vee \gamma)) \qquad \text{for } N' \text{ a measurement of } \beta, \gamma$$

$$= \mathrm{df} \ \Box_{M'}(\alpha \cdot \beta \vee \gamma)) \qquad \text{for } M' \text{ a measurement of } \alpha, \beta, \gamma$$

$$= \mathrm{df} \ \Box_{M'}((\alpha \cdot \beta) \vee (\alpha \cdot \gamma)) \qquad \text{by distributivity of } \vee \text{ and } \cdot$$

$$= \mathrm{df} \ \Box_{M''}(\alpha \cdot \beta) \vee^* (\Box_{M'''}(\alpha \cdot \gamma)) \text{ for appropriate } M'', M'''$$

$$= \mathrm{df} \ (\Box_M \alpha \cdot^* \Box_N \beta) \vee^* (\Box_M \alpha \cdot^* \Box_L \gamma).$$

their 'certain presence' or 'certain absence' after measurement. He argues that these operations among properties are represented by operations among projections or the corresponding subspace operations of Hilbert Space. The resulting 'calculus of properties' is assumed by von Neumann also to be a logic of propositions. Non-standard features of these Hilbert Space operations are supposed to show that 'quantum logic' also is non-standard.

Von Neumann's analysis is examined and re-expressed here in strictly logical terms. Although properties and wffs of the propositional logic of a theory correspond, properties are represented by modal formulae in the modal extension of this logic. An atomic property, for example that 'quantity m has value r', corresponds to the proposition $p = (m, r)$ in the logic of a theory, but it is actually represented logically by the modal wff $\Box_m p$, for when the property is 'certainly present' after measurement, $\Box_m p$ is true. Operations among properties are therefore defined as operations among modal wffs of this kind. For example, the 'conjunction' of two properties $\Box_m p$ and $\Box_n q$ is the property $\Box_{M'}(p \cdot q)$, for M' a measurement of both p and q. When there is no such measurement the operation is not defined.

Using this representation of von Neumann's 'calculus' one can see that there are important differences between the properties of classical and quantum theories. In a classical theory, each wff in LM_T has a corresponding property, and operations among properties are defined everywhere since appropriate measurements can always be found. However, in non-classical theories such as those of quantum mechanics, wffs may not be compatible and so there will be more wffs than properties, and operations among properties will not always be defined. The calculus of quantum properties is therefore only a partial algebra.

Finally, we returned to Birkhoff and von Neumann's experiment, showing that even when this example is re-stated in terms of properties rather than propositions it does not establish that operations in this calculus are non-distributive. In fact, where they exist, the operations are distributive in any theory. What first appears to be a failure of distribution is instead a case where operations cannot be defined.

Hence von Neumann's argument does establish some differences between the calculus of properties in classical and quantum theories, but it does not show even that this calculus is non-distributive. Clearly neither the experiment in Birkhoff and von Neumann's paper, nor the general analysis, succeed in showing that distribution fails in the *logic* of a quantum theory. We have no reason to suppose that the propositional logic of quantum theories, which we use to reason about microscopic things, is radically non-classical.

8 Conclusions

8.1 Return to the paradoxes

It seems worthwhile now, after the detailed discussion in the last six chapters, to return to the quantum 'paradoxes' which were discussed in chapter 1. After all, these illustrate the difficulties of understanding quantum mechanics. If the logical analysis has resolved any fundamental problems, the paradoxes should now be easier to understand. In chapter 7 we examined in detail one of these, which Birkhoff and von Neumann used in support of a non-distributive 'quantum logic'. We shall now much more briefly examine the others. I hope to show how the earlier principles of logic, measurement and probability theory can 'interpret' these quantum experiments.

8.1.1 Schrödinger's cat

Recall that this paradox centres on the use of von Neumann's 'Projection Postulate' in quantum theories. The 'thought-experiment' was described in §1.2.1. We suppose a cat is placed in a glass box, with a microscopic electrocuting device which can be triggered by a single electron if it passes through a half-silvered glass. We then suppose a single electron is fired, which will have a fifty-fifty chance of triggering the device. According to quantum theory, after the experiment but before an observation is made, the cat will be described by a superposition of 'dead' and 'alive' quantum states, which will 'reduce' to one or the other after an observation is performed. This generates two paradoxes. First, it seems that the cat is neither dead nor alive in this period after the experiment but before observation. Secondly, a seemingly non-disturbing measurement—a mere glance at the cat in the box—appears to radically alter the cat. The state of the cat 'jumps' after this simple measurement to being either quite dead, or utterly alive.

Now, according to the present views, the 'Projection Postulate' is a conditionalising rule appropriate to theories with non-bivalent states. According to this rule, we generally have to alter our description of a system when new information is received, sometimes in a very drastic way. The quantum state which describes the cat after the experiment but before observation is not an incomplete version of some complete 'picture' of the system. It tells us something about the reality, for example what probabilities are associated with the two possible outcomes of an observation. However, unlike a corresponding classical description, new information cannot simply be added to it to make a complete 'picture'. When we do discover something new, for example that the cat is alive, then we need a new description of the system, expressed by a new and different quantum state.

Before the cat is observed, we would also describe it in a classical theory by assigning probabilities to the possible outcomes of a measurement, but in a classical theory the state which is 'discovered' by glancing at the cat is also assumed to be the state of the cat before we glanced at it. The earlier statistical description is understood as an 'incomplete' version of this state. In theories without bivalent states, however, the situation is different. The superposition describing the cat before observation is, according to quantum mechanics, a full characterisation of the system, and has just the same status in the theory as the 'reduced' state of the cat after it is observed. We conclude from our knowledge of cats that if the cat was alive afterwards then it was also alive before the observation, but this is an 'extra-theoretic' conclusion which is not allowed, let alone deduced, from quantum theory. Clearly, the theory is unable, in this example, to tell us everything we want to know about the cat.

According to this view, quantum theory is indeed inadequate, or 'incomplete' in the sense that it cannot assign truth-values to all its simple propositions. Before the cat is observed after Schrödinger's experiment, the proposition 'the cat is alive' is undecided: the state of the theory which describes it does not assign a truth-value to this proposition. Thus when the observation is performed, and a truth-value is assigned to this proposition, a new state must be used. Notice, however, that quantum theory is not 'incomplete' in the sense that it fails to provide information which could be expressed by the state—it is not simply that the quantum state is contained in some wider 'picture' or valuation which does make the appropriate truth-assignments. This is because the assignment of either truth-value to this proposition cannot be consistently added to earlier truth-assignments. It is for this reason that the state describing the system must be changed after observation.

However, in quantum theories states are not uniquely associated with realities, and so this change in the state of the cat does not signal a change in the cat itself. As I stressed earlier, one must carefully distinguish between the states of our theories and the states-of-affairs which these describe. There is no unique relation between the two in quantum theories, as Schrödinger's experiment illustrates. We have here a 'jump' in our description, in our knowledge about the system, not a 'jump' in the system itself.

Thus the second part of the paradox is also resolved. Since the cat is unchanged after the experiment we do not need to explain how our glance could alter it. Measurement essentially provides information, and it is this which accounts for the change in our description of the cat. We do not need to consider additional properties of measurement, such as its disturbance of small systems. The 'jump' in state can be explained by logical principles alone.

8.1.2 EPR

The paradox of EPR is likewise easy to explain. After all, it was devised to show that disturbance interpretations are problematic, and these interpretations are rejected here. Recall that in this thought-experiment two systems interact briefly, their magnitudes becoming correlated, and then they separate. Again there are two parts to the paradox. After the interaction a measurement on one system will cause the states of both systems to 'jump', even though they are spatially far apart. Secondly, our choice of measurement on one system will determine the kind of 'jump' which occurs, even on the separated system.

Once again, the paradox disappears if we accept the Projection Postulate as essentially a conditionalising rule, and so recognise that 'jumps' in the state of a system may not correspond to physical 'jumps' in the state-of-affairs described. Measurement on one system yields information about both, by means of the correlations. When we conditionalise on this new information we are forced to change our earlier state-descriptions, since the theory is non-classical. Naturally, the kind of alteration we make in our description will depend in part on the kind of information we receive, and so we would expect that our choice of measurement affects the final state of both systems.

Once more we must be careful to distinguish the states of our theory from physical states-of-affairs. State-descriptions of non-classical theories cannot express information about every magnitude, because the elementary propositions of the theory cannot all be consistently assigned truth-values. We must therefore accept that more than one state may refer to the same reality, and that information may be 'lost' in successive descriptions. In such a theory one can only assume local consistency, i.e. that immediately succeeding descriptions do not actually conflict. One cannot generally assume the stronger condition that theories express globally consistent 'pictures' of reality, in which all propositions are eventually assigned truth-values in a consistent way.

In a sense, therefore, we agree with Einstein. The EPR experiment shows that quantum descriptions are 'complete' only in the weak sense that no more information can be consistently added to them. They are not complete in the stronger sense of finding every proposition true or false. The 'jump' in state after measurement is a result of this weakness, or 'incompleteness', in the quantum descriptions.

8.1.3 The two-slit experiment

Recall that in the two-slit experiment a source of electrons is placed on one side of a diaphragm punctured by just two slits, with a photographic plate placed on the other side (§1.2.5). Electrons are fired at the diaphragm, with 'hits' being registered as dark patches on the photographic plate. The two-slit 'paradox' arises when we compare some microscopic experiments with their classical macroscopic counterparts. In particular, we note that if just one slit is open in the electron experiment, and the other slit is closed, then the pattern on the plate indicates a concentration of hits opposite the open slit. This is just what one expects from a macroscopic experiment, for example with bullets or billiard balls being fired through a hole. However, when both slits are open, the electrons register an interference pattern on the plate. This is unlike the simple 'additive' pattern concentrated opposite each hole, which appears in the corresponding classical experiment. Furthermore, the interference pattern builds up in the microscopic experiment even when the electrons leave their source just one at a time. If any attempt is made to measure the system, for example to register their passage through the slits, then they are deflected by the measurement and the normal 'additive' pattern appears.

We saw in chapter 1 that Bohr used this example to support his early 'disturbance' interpretation of quantum mechanics, and his later more complex views on 'complementarity'. These views will be discussed again briefly in the next section. The two-slit experiment is also frequently supposed to demonstrate a logical paradox. It is often argued, in a way that is analogous to von Neumann's argument, that the distributive law breaks down in quantum theories. Alternatively, in a similar way, it is claimed that a classical law of probabilities fails. Both these claims can be rejected.

One can label the two slits on the diaphragm, A and A′, and can label some point on the photographic plate B. When the experiment is performed, simple propositions a, a′ and b can be associated with these positions. For example, a is the proposition 'the electron is at position A'. According to the argument, we consider three different experiments, one in which slit A alone is open, another with only A′, and a third in which both slits are open. In each case, it is argued, a different complex proposition can be associated with the point B and hence with proposition b. Where slit A alone is open, the point B corresponds not only to the proposition b, 'the electron arrived at B', but also to the conjunction (b · a), 'the electron arrived at B, after passing through slit A'. Similarly, in the second experiment, where only A′ is open, B is associated with (b · a′), 'the electron arrived at B, after passing through A′'. In the third case, where both slits are open, B is supposed to correspond to b · (a ∨ a′), 'the electron arrived at B, after passing through A or through A′'. The argument proceeds by noting that it is possible to have a 'hit' at point B in the first experiment, but a blank at this position in the second and third. Thus, it is argued, (b · a) may be true, while (b · a′) and b · (a ∨ a′) are each false. It follows from the definition of disjunction that in this case, (b · a) ∨ (b · a′) is true, while b · (a ∨ a′) is false, and this is supposed to indicate a failure of distribution.

The argument about probabilities is very similar. Here, the probabilities associated with position B in each experiment are compared. It is assumed that distribution holds. As in the earlier argument, when both slits are open B is associated with the proposition b · (a ∨ a′), and since distribution is assumed, also with (b · a) ∨ (b · a′) (the distributive identities are D1 and D2 of §2.2.1). By the 'Law of Total Probability' (derived from Kolmogorov's fifth axiom, §5.1.1), $prob(\alpha \vee \beta) = prob(\alpha) + prob(\beta) - prob(\alpha \cdot \beta)$. Since a and a′ are negations, with zero joint probability, this law tells us that $prob((b \cdot a) \vee (b \cdot a')) = prob(b \cdot a) + prob(b \cdot a')$. However, the probabilities on the right-hand side are supposed to be given by the first two experiments—the first where only A is open, and the second with only A′. Therefore one expects the ordinary 'additive' pattern to appear in the two-slit case, just as one does in macroscopic experiments. Since this is not actually observed, it is argued that the law of total probability fails. According to this view, the Kolmogorov axioms of probability must be revised in quantum theory.

One should note at once that both these arguments show confusion about how the logic of a theory is actually applied in a specific case. Strictly speaking, logical connectives are properly defined only among propositions which are used by a theory to describe exactly the same physical reality, and hence they should be describing a system at the same instant in time. In the analyses above of the two-slit experiment this condition is not met, since proposition b, for example, concerns a different time from propositions a and a′. The arrival at A or A′ takes place before the arrival at the photographic plate. Thus, strictly speaking, the complex formulae associated with B are not well defined. Similarly, the three different complex formulae (b · a), (b · a′) and b · (a ∨ a′), are all descriptions from different experiments, and so, for example, the disjunction ((b · a) ∨ (b · a′)) is also not properly defined.

However, even if the expressions were well defined in some logic describing the experiments, the argument fails for reasons similar to those discussed already for von Neumann's experiment. In both cases the proposition b is assumed to be equivalent to the conjunction b · (a ∨ a′). Hence in both cases (a ∨ a′) is assumed to be logically true. 'Before

arriving at the plate, the electron passed through either A or through A''
is supposed to be self-evident, and so it is assumed to be true in every
valuation of the theory, but this assumption is confused. One can agree
that this disjunction can never actually be false, since a and a' are indeed
negations in the logic of the theory, but this is not to say that it is logically
true. In a non-classical theory such as quantum mechanics, where states
do not decide all wffs of the theory, this disjunction may be undecided.
For instance, a state describing the momentum of the system at the
diaphragm will not also be able to describe its position, and so in such a
state both a and a' will be undecided, as will a $\vee$ a'. Since a $\vee$ a' is not a
tautology, it follows that b $\cdot$ (a $\vee$ a') is not equivalent to b, and so both the
arguments fail to establish their conclusions.

The same remarks apply to the conjunctions b $\cdot$ a and b $\cdot$ a' in the other
experiments. Proposition b is not equivalent to these conjunctions, for
proposition a is not *logically true* even when slit A is the only open
aperture in the diaphragm. It may seem that 'The electron arrived at B
after passing through A' is just the same as 'The electron arrived at B' in
this experiment, but logically these propositions are not equivalent. The
theory has incompatible magnitudes and so there will be states describing
the experiment in which proposition a is undecided.

Thus there is no logical paradox in the two-slit experiment, nor is any
Kolmogorov axiom violated. What we have here is a non-classical
description. Our choice of magnitudes and value-sets in quantum
mechanics determines that propositions a and a' are relative complements in E_T, and so are negations in the logic LM_T of the theory. The
descriptions given by T must be consistent with these relations and so the
disjunction of negations a $\vee$ a' is never false. In this sense, it is a
dichotomy in any description given by T. But this does not ensure that the
disjunction a $\vee$ a' is an 'absolute', or 'real' dichotomy, in the sense that it
is also logically true, i.e. true in every state-description.

One can agree with Bohr that the loss of the interference pattern
during measurement is indeed a consequence of physical disturbance. He
has shown, in detailed analyses, that measurements do affect this
experiment. For example, by placing counters at the slits to register the
passage of electrons, we deflect them from their original path and so
destroy the very phenomenon we wished to measure. Similarly, he shows
that other measurements of other magnitudes do disturb this experiment.

Yet one need not agree with Bohr that the two-slit experiment shows
that measurement disturbance in some sense causes the incompatibility of
magnitudes in quantum mechanics. According to the present view,
incompatibility is a logical property arising because non-classical theories
use systems of elementary descriptions which cannot all be assigned truth-
values in a consistent way. Any such theories have non-bivalent states,
and we have shown that quantum-type peculiarities will arise in their

probabilities. However, no reference is made to physical properties of measurement in this analysis. Since we have no reason to suppose that the relations among simple propositions depend on these physical properties, we have no reason to conclude that physical disturbance in any sense 'causes' the non-classical peculiarities.

Of course, this is not to say that measurement disturbance is irrelevant to the discussion of quantum mechanics. If one could easily measure incompatible magnitudes, assigning accurate values to both at once, then perhaps our quantum descriptions would long ago have been considered unsatisfactory. It is probably because of the physically disturbing properties of microscopic measurement that we have had to live so long with a non-classical quantum theory.

This example shows just how much and just how little the logical analysis of quantum theories can explain. We see that contrary to many claims no law of logic or of Kolmogorov probabilities is violated in the two-slit experiment. In this sense, a 'paradox' is resolved. Our means of describing the experiment and our laws of reasoning are essentially those of ordinary logic, but we recognise that our theory uses states which are not bivalent, and in this sense it gives no globally consistent 'picture' of the reality. Thus another problem remains: we know that our descriptions are not quite adequate, but this logical analysis does not provide a more adequate description of the system. It does not provide an alternative theory. The logical analysis tells us how to understand our own theory and our own rules of reason, but it does not show us how to understand and more fully describe microphysical reality.

8.2 Bohr and Einstein

It is now time to consider the wider issues raised by this analysis, and to compare this interpretation with the views of Bohr and Einstein, very briefly presented in chapter 1. Recall, from §1.2, that their dispute resulted only in a fundamental and unresolved difference of opinion. Bohr argued that quantum mechanics was a great advance on classical theory, introducing new 'complementary' concepts necessary to describe microphysical realities. Einstein, on the other hand, always argued that quantum theory was essentially inadequate and its statistical descriptions 'incomplete'. It is impossible in the space available to do justice to the complexities of either view. However, even a brief comparison between Bohr's and Einstein's views may make the present interpretation of quantum theory more clear. We start with a brief review of this interpretation.

8.2.1 *The logical interpretation*

Here it has been argued that the peculiarities of quantum theories are explained by their lack of bivalent states. We showed that in any such theory states are essentially statistical, in the sense that they are not bivalent measures over a unique phase space. Conditional probabilities in these theories are not traditional 'ratio' conditionals, so when new information is established by measurement, providing new conditions, earlier probability assignments may be radically revised. Since there is no way that all the simple descriptions can consistently be combined in a state, previous information may be 'lost' when we find out something new.

Since the oddities of quantum theories arise from a lack of bivalent states, they stem essentially from the structure of the elementary sets of propositions. Generally, the simple propositions of set E_T used by a quantum theory T cannot all be consistently assigned to the two truth-values and for this reason states of the theory will not be bivalent. One must conclude, therefore, that it is the choice of magnitudes used, their values and the laws of the theory which cause the peculiar features of quantum mechanics, for it is these that determine the structure of the elementary system E_T.

This logical analysis relied only on minimal assumptions about the nature of mechanical theories and made no assumptions at all about the procedural nature of measurement or the physical nature of the realities described. We were concerned throughout only with the logical properties of descriptive theories and not with the nature of description itself or the nature of reality. For instance, it was assumed that mechanical propositions have the form (m, Δ), generated from a fundamental set of magnitudes M_T and their sets of values. It was also assumed that set relations among the values generate corresponding relations among the propositions. A state was defined logically as a maximal valuation of propositions, and measurement too was characterised solely in logical terms, as some kind of process which has the essential property of producing truth-value assignments to simple propositions. The preceding analysis shows that these assumptions alone can account for the peculiar features of quantum mechanics. The peculiarities arise from the logical structure of the theories, in particular from their lack of bivalent states.

Since we can explain the quantum peculiarities solely in logical terms, as arising solely from the structure of the theories themselves, it is clear that we have no reason to introduce additional assumptions about the physical nature of measurement or the realities which quantum theories describe. It may indeed be the case that both have extremely interesting and unique properties because of their extremely small size, but we need not appeal to these properties in order to understand the quantum statistics. In this sense, we can regard quantum theories as being 'idealised' theories in exactly the sense that classical theories are idealised. The physical inaccuracies and physical problems of measurement are not taken into account in the mathematical structure of the theory itself.

8.2.2 *Bohr*

Bohr and Einstein approached the problems of quantum theory with strong and opposing philosophical prejudices which often obscured their arguments and which prevented them from finding any common ground. Bohr, for example, recognised that our use of language lay close to the heart of the problem. To this extent his view agrees with ours. Our choice of terms, in particular our choice of magnitudes, value-sets and laws will determine whether a theory is classical or non-classical, and hence whether it has the quantum peculiarities. Bohr was acutely aware of the ways in which language can limit our understanding and our descriptions of things, and he considered us bound to the use of certain terms. For example, Bohr wrote (Schilpp 1969 p 210):

> ... it is decisive to recognise that, however far the phenomena transcend the scope of classical physical explanation, the account of all evidence must be expressed in classical terms.

Bohr argues further that it is our use of macroscopic measuring instruments which limits us to classical concepts. In his early writing the influence of measuring instruments was understood as essentially physical. Our observations physically alter or destroy the realities observed. However, as we saw in chapter 1, this simple 'disturbance' view runs into problems when confronted with such cases as Schrödinger's cat, or the EPR experiment. After the EPR paper Bohr claimed that our language itself depends on our use of measuring instruments, since these (Schilpp 1969 p 210):

> serve to define the very conditions under which the phenomena appear.

As we saw in §1.2.4, this leads him to claim, in response to EPR, that even where no physical disturbance is caused by measurement, there can still be (Bohr 1935 p 700):

> ... an influence on the very conditions which define the possible types of predictions regarding the future behaviour of the system.

In the same way, Bohr explains the occurrence of incompatible magnitudes in terms of the instruments and procedures which measure them. It is because the physical procedures are incompatible that these magnitudes cannot be assigned precise values simultaneously. After the

EPR paper he cannot claim it is simply a physical disturbance which prevents us from finding out these precise values. Instead, he claims that the measuring instruments are somehow involved in the very definitions of our terms, so that the physical impossibility of measuring them makes the assignment of precise values to incompatible magnitudes meaningless. As we saw in §1.2.5, this is how Bohr resolves the two-slit paradox. Because one cannot meaningfully assign precise values to position and momentum magnitudes, there is no such thing as the 'trajectory' of an electron in this experiment.

Although there is much in Bohr's analysis which is sympathetic to the present view, one cannot agree with his emphasis on measuring procedures. One can agree that our descriptions are limited by the terms we use, and that we perhaps use forms of description which are unsuited to microscopic systems, but he does not justify the claim that measuring instruments are fundamental to our choice of terms. In classical theories one assumes that magnitudes have precise values, even though these might never possibly be measured. Why should this not be the case in quantum theories? Also, his reply to EPR is surely dogmatic and far from satisfying. Bohr seems unwilling to suppose that anything other than measuring procedures might explain the incompatibility of magnitudes. His philosophical prejudices are particularly clear when he considers the future of quantum theory. Because magnitudes of quantum theory are incompatible, Bohr claims that reality itself is in some sense 'complementary'. This leads him to claim that any future theory describing microscopic reality will have the same peculiarities as quantum mechanics. According to Bohr (1935 p 700), only the use of incompatible magnitudes will 'provide room for new physical laws'. He always maintained that quantum mechanics marks an advance and indicates the form of future physical theories.

This conclusion is certainly rejected here. The logical analysis showed that in non-classical theories information is lost, and state descriptions must be changed when measurements introduce new information about a system. It was also shown that in classical theories this is never the case, since where states are bivalent, new information can always be consistently combined with previous descriptions. This difference between the theories provides a criterion for calling classical theories the more successful. Classical theories provide globally consistent descriptions of reality. Although one cannot be sure that a classical theory of microscopic systems is possible, there is no reason to suppose *a priori* that it is not. If such a theory were developed, with experimental success, it would be preferable to non-classical quantum mechanics.

8.2.3 *Einstein*

Einstein's arguments show a different philosophical prejudice. Recall that his argument in the EPR paper for the incompleteness of quantum descriptions is based on ontological premises. He defines a *complete* theory as one in which (Einstein, Podolsky and Rosen 1935)

> every element of reality has a counterpart in the physical theory.

An 'element of reality' is defined as any quantity whose value can at any time be predicted with certainty, i.e. with probability equal to unity.

These premises appear to state a crude naive realism. The terms of our theories correspond to real 'things'. According to Einstein, if we can assign probability 1 to a proposition '*m* has value *r*', then there must be some real 'element' in reality corresponding to *m*. Einstein's argument for the incompleteness of quantum mechanics is based on this strong ontological position. Because quantum magnitudes are incompatible, they cannot all be assigned precise values with certainty in a single state, and so quantum states do not describe every 'element of reality'. They are therefore 'incomplete'.

Einstein has been rightly criticised, by Bohr among others, for these dogmatic principles, and yet his discontent with quantum theory does not seem to rely necessarily on these assumptions. If one simply ignores his ontology, making no assumptions about reality, much of Einstein's view agrees with the present analysis.

The essential point argued by Einstein was that quantum states express descriptions which are incomplete. This is entirely in accordance with the present view. States are maximal valuations of a theory, expressing its 'fullest' possible descriptions, but in some theories, those which have been called non-classical, states will not be bivalent. In such theories, including those of quantum mechanics, it is impossible to assign truth-values to every proposition in a consistent way. In this sense the descriptions of quantum theories are clearly incomplete.

An argument similar to Einstein's can therefore be re-expressed without appeal to any ontological assumptions. Instead, we can say that a description is *complete* only if it decides every simple proposition of a theory. Likewise, a theory is complete only if all its states are, i.e. only if it has bivalent states. We have seen that only complete theories use classical probabilities of the familiar kind. In non-classical theories, whose states are not bivalent, states will be irreducibly statistical and the probabilities assigned in these theories will have features similar to those of quantum theory. We can say, therefore, that these features characterise a theory which is incomplete.

Since Einstein assumed that theoretical terms correspond to 'elements of reality', he concluded that a theory whose states predict a value for each magnitude with certainty gives a better description of reality than a theory which does not. He thus considered classical theories to be better than non-classical theories. Here too we can agree with Einstein's general conclusion without accepting his ontology. A non-classical theory can give only a locally consistent description of reality. Its states are not globally consistent and so information is often lost after measurement. A theory with bivalent states, however, has no information loss after measurement. In a classical theory all descriptions can be assigned probability 1 or 0 in a state, while in non-classical theories only statistical predictions are generally possible. In these respects one can clearly regard classical theories as superior to non-classical ones. If two theories accord equally well with our observation we would choose a classical rather than a non-classical theory. In this sense classical theories are the theories of the future.

It is interesting to ask whether the present quantum theory can be extended in some way to provide a full and classical description of reality. This is to ask whether the elementary system E_T of simple propositions used by a quantum theory can be augmented so that the present relations are preserved, but bivalent valuations are admitted. Kochen and Specker (1967) have published a result purporting to show that this is impossible. They argue that, in general, the elementary set E_T of a quantum theory cannot be embedded onto the two-element Boolean algebra Z_2, and so bivalent assignments to an augmented elementary set will not be possible. If this conclusion is correct, then one must conclude that quantum mechanics cannot simply be extended into a classical theory. Instead, it must be modified in some rather drastic way, or replaced altogether by a theory with a new system of elementary descriptions.

8.3 In conclusion

8.3.1 Logical remarks

We now see what logical lessons can be learned from this analysis of mechanics.

First, of course, it makes the nature and role of logic in these theories more clear. According to this view, the propositional logic of a mechanical theory is a slightly generalised version of classical bivalent logic. The logic allows truth-value 'gaps', and so is founded on three-valued valuations. In such a system, inference has its familiar classical properties, and so too have disjunction, conjunction and the bicon-ditional, while negation is also essentially classical but now restricted in scope. One can make distinctions in this three-valued system which are impossible in a bivalent logic: for example, we can distinguish between negation and denial, since to say that a proposition is not true in a description is now distinguished from saying that it is actually false. We saw that this generalisation of bivalent logic introduces a natural sense of relevance into the system, so that negation has its strong classical properties only among relevant formulae. Algebraically, this logic is represented by a distributive lattice in which every element has a relative orthocomplement.

Hence we do not agree with those who argue that quantum theories use a non-distributive 'quantum logic' to describe microphysical systems. This would be a system with radically non-classical inference. After examining the arguments we concluded that von Neumann does indeed show a difference between the properties of classical and non-classical theories, in the special sense he gives this term. The calculus of properties in a classical theory is an algebra, in which operations between properties are always defined. In quantum theories this is merely a partial algebra, where the 'simultaneous measurability' of magnitudes must be taken into account. Close analysis shows, however, that von Neumann's calculus of properties is not a propositional logic, nor even a modal logic of a theory.

The modal logic of mechanical theories was introduced as a logic in which some predictions can be expressed. One can say in this logic not only that descriptions are true, false or undecided, but also that they are certain to be true after measurement, or will possibly be true after measurement. The logic was generated using a new kind of definition for modal operators.

The modalities discussed here and the whole analysis of probabilities depend on a thorough investigation of the logical properties of measurement and its role in description. Measurements are essentially procedures which establish truth-values of propositions. No physical properties of measurement are considered. We saw that in some theories new information may cause us to radically alter our descriptions of a system, and so measurement may play a vital role in changing our descriptions. It is simply the logical properties of measurement, the fact that new truth-values are introduced, which give it this crucial role.

One important lesson to be learnt from this analysis of measurement is that we must distinguish our descriptions of reality from realities themselves. The states of a theory are maximal valuations—the 'fullest' descriptions a theory can give of reality. These are essentially different from the states-of-affairs, the physical realities which are described.

In classical theories it has not seemed important to make this obvious distinction. In these theories only one state can describe each state-of-affairs and so we have come to assume that reality exactly resembles our theoretical descriptions. Even in this classical case our assumption is fundamentally unjustified, since the mere fact that our description is unique will not ensure that it exactly 'mirrors' reality. Clearly, we can describe only our own perception of a system, while reality in itself remains essentially out of reach. In spite of this, however, the unique relation between states and states-of-affairs in classical theories does at least mean that we can use the single term 'state' for both, without great ambiguity or paradox. For example, the unique correspondence assures us that any change in the state describing a system must indicate a corresponding change in the state-of-affairs.

In non-classical theories, the situation is very different, and it becomes vital that we do distinguish carefully between our descriptions of reality and reality itself. In these theories there is no unique relation between states of a theory and the states-of-affairs described. No state can make truth-value assignments to every proposition of the theory and so more than one state may be used to describe the same reality. Whereas in classical theories we may find it convenient to forget that states are no more than descriptions of reality, in non-classical theories such as quantum mechanics we cannot ignore this fact. If we use the single term 'state' for both our descriptions and the reality described, we are led at once into paradox. For example, since states do not uniquely characterise states-of-affairs, sudden changes in state may not describe sudden changes in reality. We have seen examples of this situation in the 'paradox' of Schrödinger's cat, or the EPR experiment. Quantum states may be altered, while the system they describe remains the same.

Thus quantum theories provide a 'weak' description of reality, in which the maximal descriptions are not uniquely associated with states-of-affairs. This accounts for many of the peculiarities of quantum probabilities. The statistical nature of states, the 'jumps' in state after measurement, the peculiar sequential probabilities which depend on the order in which measurements are performed, are all features of a theory in which states cannot uniquely characterise the states-of-affairs which they describe. These are features of a theory which gives only weak descriptions of reality.

Perhaps the most important part of this analysis is the new logical foundation it gives for probabilities. Probabilities are measures of sets of states, i.e. of maximal valuations of the theory. The probability of a wff is a measure of the set of 'possible' states in which this wff is true. In classical theories, this coincides with our normal understanding of probability statements and we showed that in this case probabilities can be represented in the usual way, as measures over a classical phase space. However, in general, probabilities are strongly conditional. Different sets of states will count as 'possible' according to different initial conditions, and so we generally require many different probability spaces to generate all the assignments of a theory. We saw that this can account for all the striking features of the quantum statistics.

Thus the difference between classical and quantum theories is not a difference in the laws of logic which are used, nor in the fundamental properties of their probability measures. All the differences stem from the different properties of their valuations, and hence of their states. The propositional logic used by these theories is the same, and their states and valuations generate probability assignments in exactly the same way, but the special properties of classical states mean that classical probabilities, and corresponding modalities, are also special. They are conditional only in the weak traditional sense, and they can all be defined over a single phase space of states. Quantum theories are not classical, however, and so their probabilities do not have these special properties. Instead, they have the strongly conditional probabilities and non-trivial modalities, which characterise those 'weak' non-classical theories which lack bivalent states.

8.3.2 Conclusion

If this analysis is adopted we see the peculiarities of quantum mechanics as features of our own description rather than features of the microscopic world. It is the states of our theory—our descriptions of reality—which are fundamentally statistical, not the reality they describe. Similarly, it is our descriptions of a system which 'jump' after measurement, not the system itself.

Although this general view of quantum mechanics is not new, it is supported here by a new analysis of the foundations of the theory. One can argue, in general terms, that quantum mechanics is inadequate, leaving us ignorant about microscopic reality, and one can claim that this ignorance explains the quantum peculiarities, but this is not a fully fledged interpretation of the theory until it is supported by rigorous analysis. We need to know in exactly what sense we are ignorant, and exactly how this generates the features of the theory that we wish to explain. The preceding chapters have been devoted to just this kind of analysis. We have seen in what sense non-classical theories are inadequate, and we have seen how probabilities with the quantum features arise in these theories.

This logical interpretation of quantum mechanics can claim a measure of sense and reason. It is surely sensible to see the quantum peculiarities as products of weak description rather than as incomprehensible features of the world. It is surely more reasonable to suppose that our theory is inadequate than to argue that reality itself is bizarre. Furthermore, by adopting this view we are left with our usual hopes and metaphysical assumptions. We hope that reality is open to our understanding and for the purposes of doing science we assume that it is so. There is no reason to suppose that subatomic reality lies beyond our comprehension. We should accept quantum mechanics as the most successful theory we have at present, while setting out to develop a new and better description of reality. It is in the search for a new bivalent description of subatomic systems that advance in the subject lies.

References

Accardi L ed 1982 *Conference on Quantum Probability* (Rome: Springer)

Bauer H 1972 *Probability Theory and Elements of Measure Theory* (London: Holt, Rinehardt and Winston)

Bell J S 1964 *On the Einstein–Podolsky–Rosen Paradox* Physics **1** 195–200

Bell J L and Slomsen A B 1974 *Models and Ultraproducts: An Introduction* (Amsterdam: North-Holland)

Birkhoff G 1967 *Lattice Theory* 3rd edn (Providence RI: American Mathematical Society)

Birkhoff G and von Neumann J 1936 *Ann. Math.* **37** 823–43. Reprinted in Hooker ed (1975)

Bohm D and Aharonov Y 1957 *Phys. Rev.* **108** 1070–6

Bohr N 1935 *Phys. Rev.* **48** 696–702

—— 1958 *Atomic Physics and Human Knowledge* (New York: Wiley)

Borokowski L 1970 ed *Jan Lukasiewicz: Selected Works* (Amsterdam: North-Holland)

Bub J 1974 *The Interpretation of Quantum Mechanics* (Dordrecht: Reidel)

—— 1977 *J. Philos. Logic* **6** 381–90

—— 1979 *The Measurement Problem of Quantum Mechanics* in *Problems in the Foundations of Physics* (Bologna: Società Italiana di Fisica)

Chari C T K 1971 *Synthese* **22** 438–47

Colodny R G 1965 ed *Beyond the Edge of Certainty* (Englewood Cliffs NJ: Prentice-Hall)

—— 1972 ed *Paradigms and Paradoxes* (Pittsburgh: University of Pittsburgh Press)

Cresswell M J 1975 *Frames and Models in Modal Logic* in *Algebra and Logic* ed J N Crossley (Berlin: Springer)

Einstein A, Podolsky B and Rosen N 1935 *Phys. Rev.* **47** 777–80

Gemignani M 1970 *Finite Probability* (London: Addison-Wesley)

Gillespie D T 1973 *A Quantum Mechanics Primer* (London: Intertext)

Gratzer G 1971 *Lattice Theory, First Concepts and Distributive Lattices* (San Francisco: W H Freeman)

Halmos P 1951 *Introduction to Hilbert Space and the Theory of Spectral Multiplicity* (New York: Chelsea)

Hardegree G 1977 *Relative Compatibility in Conventional Quantum Mechanics* in *Foundations of Physics* **7**

Hooker C A 1972 *The Nature of Quantum Mechanical Reality: Einstein versus Bohr* in *Paradigms and Paradoxes* ed R G Colodny (Pittsburgh: University of Pittsburgh Press)

—— ed 1975 *The Logico-Algebraic Approach to Quantum Mechanics* vol 1 (Dordrecht: Reidel)

Hughes G E and Cresswell M J 1972 *An Introduction to Modal Logic* (London: Methuen)

Hughes G E and Londey D G 1965 *The Elements of Formal Logic* (London: Methuen)

Jammer M 1974 *The Philosophy of Quantum Mechanics in Historical Perspective* (New York: Wiley)

Jauch J M 1968 *Foundations of Quantum Mechanics* (London: Addison-Wesley)

Kochen S and Specker E P 1967 *J. Math. Mech.* **17** 59–67. Reprinted in Hooker ed (1975)

Kolmogorov A N 1956 *Foundations of the Theory of Probability* (Engl. Transl. 2nd edn) (New York: Chelsea)

Kripke S A 1975 *J. Philos.* **19** 690–716

Lüders G 1951 *Ann. Phys., Lpz.* **8** 322

McCall S ed 1967 *Polish Logic 1920–1939* (Oxford: Oxford University Press)

Mackey G W 1963 *The Mathematical Foundation of Quantum Mechanics* (New York: Benjamin)

Mendelsohn E 1964 *Introduction to Mathematical Logic* (New York: van Nostrand)

von Neumann J 1955 *Mathematical Foundations of Quantum Mechanics* (Princeton: Princeton University Press)

Rasiowa H 1974 *An Algebraic Approach to Non-Classical Logics* (Amsterdam: North-Holland)

Reichenbach H 1944 *Philosophic Foundations of Quantum Mechanics* (Berkeley and Los Angeles: University of California Press)

Renyi A 1970 *Foundations of Probability* (San Francisco: Holden-Day)

Rescher N 1969 *Many-Valued Logic* (New York: McGraw-Hill)

Schilpp P A ed 1969 *Albert Einstein: Philosopher-Scientist* (La Salle Ill: Open Court)

Segerberg K 1965 *Theoria* **33** 53–71

Shively J 1980 *Quantum Logic Calisthenics* in *Studies in The Foundations of Quantum Mechanics* ed P Suppes (New York: Philosophy of Science Association)

Sikorski R 1969 *Boolean Algebras* 3rd edn (New York: Springer)

Stalnaker R 1968 *A Theory of Conditionals* in *Studies in Logical Theory* ed N Rescher (Oxford: Oxford University Press)

Suppes P 1966 *Philos. Sci.* **33** 15–26

——— ed 1980 *Studies in the Foundations of Quantum Mechanics* (New York: Philosophy of Science Association)

Wajsberg M 1930 *Axiomatisation of the 3-valued Propositional Calculus* translated and reprinted in McCall S ed 1967 *Polish Logic 1920–1939* (Oxford: Oxford University Press)

Wightman A S 1981 ed *Berlin International Conference on Mathematical Physics* (Berlin: Springer)

Index

Index of Symbols